Student's name: _____  Assignment date: _____

# H☺ Math Chess
## Learning Centre

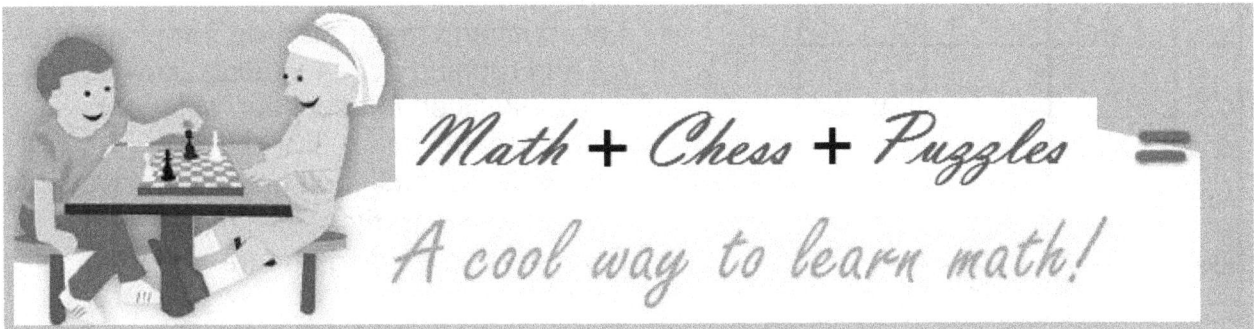

*Math + Chess + Puzzles =*
*A cool way to learn math!*

# Own a unique, math-focused, global tutoring franchise

| About H☺ Math Chess Learning Center | |
|---|---|

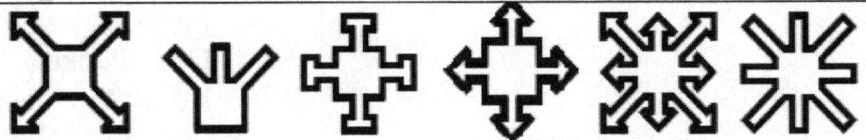

Gameplay has been demonstrated to be an effective way of teaching math. Ho Math Chess invented math, chess, and puzzle-integrated math workbooks aimed at improving children's abilities in basic computation and developing their problem-solving and creativity in a fun and inspiring learning environment.

If you have any comments or problems, please contact Ho Math Chess at fho1928@gmail.com.

Student's name: _____ Assignment date: _____

# Highlights

**Proprietary Frankho ChessDoku**

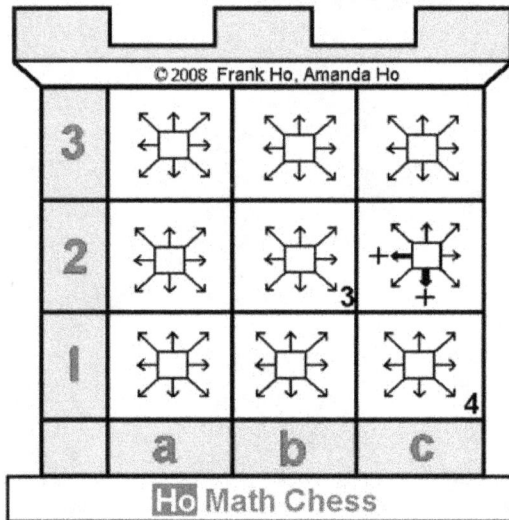

© 2008 Frank Ho, Amanda Ho

Ho Math Chess

- Proprietary copyrighted and innovative math, chess, and puzzle-integrated workbooks emphasizing analytical skills development
- Graduated puzzle difficulty builds confidence in a fun learning environment.
- Unique chess symbol and flat chess set inventions
- Low cost and faster start-up franchise (can be up and running in 2- 4 weeks) Low–cost franchise listed in education category – entrepreneur.
- Single unit or multi-unit operations available.
- The exclusive territory assigned.
- Free start-up training and free ongoing support.
- Teaching material for Pre-K to Grade 7 (ages 4 to 13) math in various workbooks: School math, Review, Math Contest, Private or enrichment math entrance exams, and SSAT.
- No inventory to stock. Low overhead and operating expenses. Simple to run with minimum staff required.
- Ho Math Chess was established in 1995, based out of Vancouver, Canada.
- Includes award-winning workbook Frankho ChessDoku
- Ho Math Chess teaching methodology has been published in a peer-reviewed math journal.

For more information, call Frank Ho (604) 263-4321, mathandchess@telus.net, www.mathandchess.com

Student's name: _____ Assignment date: _____

## Table of Contents

Student's name: _____ Assignment date: _____

Student's name: _____ Assignment date: _____

## Preface

Many people were curious about how I created so many unique workbooks. People were also wondered how Ho Math Chess is different from other similar learning centres.

There are so many different types of math, chess, and puzzles integrated problems in our various workbooks, and the majority of our students or parents do not have the chance to get access to all of our workbooks, so they do not have an idea on how these puzzles were placed in each of our workbooks. When we first created these puzzles, we did not have a blueprint to follow because they did not exist before we created them.

Many parents do not know how to teach their children with our workbooks in their hands because they had never seen this kind of arithmetic worksheets before.

So, I decided to write a book to address the above two problems, one is to describe how I created worksheets in general, and the other is to show how and why Ho Math Chess sets itself apart from others.

I sincerely would like to thank you, my wife, Amanda, who had gone along with my idea in jointly created many of our workbooks. I would like to thank my students who had given us feedback during my writing many workbooks. I wrote a few worksheets for some of my students and tried on them, and if they did not work out effectively, then I revised. Without these students' input, some of my workbooks would not have been materialized.

Finally, I hope you will enjoy using all of our workbooks, and a list of all of them for franchising is included at the end of this book.

Frank Ho

August 2015

Student's name: _____ Assignment date: _____

## Part 1 How I created Ho Math Chess Unique workbooks?

Since 1995 I have been creating new arithmetic worksheets based on one idea: the worksheets must be fun to work with and train students' computation ability, logic thinking, and creative minds. First, I invented the flat chess pieces to have chess moves carved on each chess piece's flat surface. The possible move directions are expressed by using the concept of geometry lines and line segments. Using our flat chess set, students could play chess almost immediately by just following the "what you see is what you move" approach by looking at each chess piece's lines. This chess set later becomes command language in our workbooks to guide students on how to solve a puzzle-like problem. Most of the time, students are required to write out problems instead of just answers. This extra effort of writing problems is to reinforce their memory about basic number facts. For example, in most regular arithmetic workbooks, students will just write 5 for the pre-written problem 2 + 3 = ? but in our workbooks, students need to write 2 + 3 = 5 after "playing" a puzzle integrated problem.

While I was creating these problems, sometimes I had some old soft music played in the background to get myself into a mellow and nostalgic mood and creative mind. At that point, I felt the enormous emotion coming out from myself and had a desire to create something new or out of this world.

The other day, I listened to a Chinese singer, Teresa Teng songs, who died 20 years ago and thought about creating a type of arithmetic worksheets to reflect her famous song "Sweet Honey" (https://www.youtube.com/watch?v=9fJrJ3H2MDE). I have not created it yet, but it is still in my mind. I saw children like robot class, so I created math worksheets related to the robot. I noticed that the Chinese Ba-Gua (Eight Diagrams) might have something to do with the queen's eight moves, so I created a type of arithmetic worksheet related to Ba-Gua. There were also times when I was teaching my students, and they had a hard time understanding some math concepts, so I started to think about how I could create computation worksheets to reflect the meaning of concepts. I was reading an article about smart appliances or devices. Then, an idea crossed my mind that was it possible that I could create some intelligent worksheets that will allow students to pick up the operator they wish instead of always being restricted on what they had to do? Because of this thinking, I created a type of worksheets that would allow students to pick either + or – at their choices.

I feel lucky that I could speak, read, and write Chinese to read many publications from China. China is one of the tops ranked math contest countries has taught me how they trained students. I was able to see the similarities or dissimilarities of abilities and attitudes between teachers, students, cultures, parents in different countries. Hence, I benefit from these observations and experiences tremendously when researching fun arithmetic worksheets.

Student's name: _____ Assignment date: _____

Many of my worksheets actually were trying to convey my ideas of teaching some math concepts. The above described how I created math worksheets.

When I look at these worksheets that I created in the past, I realize I would not just sit down and start to create them because the old worksheets were created at the time and environment that offered different scenery or mood of creativity.

I feel my most significant contribution has been to demonstrate that arithmetic basic number facts worksheets could also be designed to be used in exploratory nature, instead of just a plain old top-down or left to the right format. Our worksheets train students to think more and to think in multi-angle, not just one track of mind when trying to solve problems.

Creating innovative and creative worksheets could become a professional and creative job; they could be created as a very organized, structured, and powerful worksheets.
Ho Math Chess is the first organization that has created worksheets to integrate chess and puzzles into math. Students could work on them while many researchers still talk about how good the idea is to use chess or puzzles as teaching aids use them as projects.

After combing through all of my worksheets, I have classified them into six differential categories, and I gave one example for each category. These categories were not in my mind when I first started creating, but there were the end products.

Student's name: _____ Assignment date: _____

## Ho Math Chess founder Frank Ho

Frank Ho is a Canadian professional math teacher and Ho Math Chess Tutor Franchise Centre based in Canada. He has bachelor's degrees in Statistics and Computer Science and a master's degree in Statistics. He worked as a Statistical Consultant at the University of Utah, USA and the University of British Columbia, Vancouver, Canada, for almost 20 years before founding the Vancouver Ho Math Chess in Canada. Ho Math Chess Tutor Franchise Centre has a unique teaching methodology and invented a world-first innovative math workbook that integrates math, Sudoku, and chess. Frank Ho also invented the flat chess set using Geometry Chess Symbols, the core technology Symbolic Command Language (SCL) to create so many innovative and fun arithmetic workbooks.

Amanda Ho has a bachelor's degree in Science with over ten years of tutoring math experience from kindergarten to grade 12 in Canada. She has co-authored over 30 math, Sudoku, chess and IQ puzzles integrated workbooks.

Student's name: _____ Assignment date: _____

**Symbolic Chess Language (SCL)**

Ho Math Chess founder Frank Ho invents the following copyrighted Geometry Chess Symbols. They are being used as an innovative way of linking chess and math to create math and puzzle workbooks other than being used as a teaching chess set. These symbols are static, and they can only become lively commanders when Ho Math Chess used these symbols to have invented and created the arithmetic command language called Symbolic Chess Language (SCL). By using these SCL commands, students can perform arithmetic operations along with chess moves and puzzles.

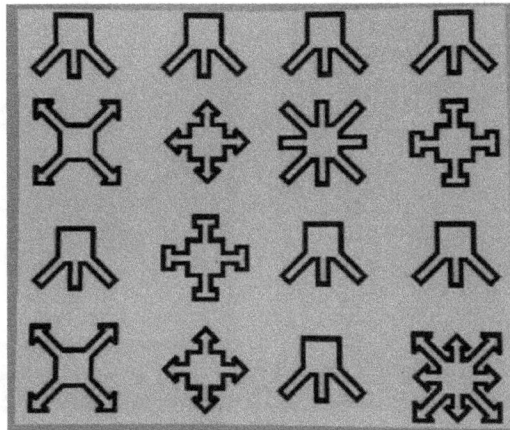

The following is the image of the Ho Math Chess Teaching set, details see http://www.mathandchess.com/chessset.html

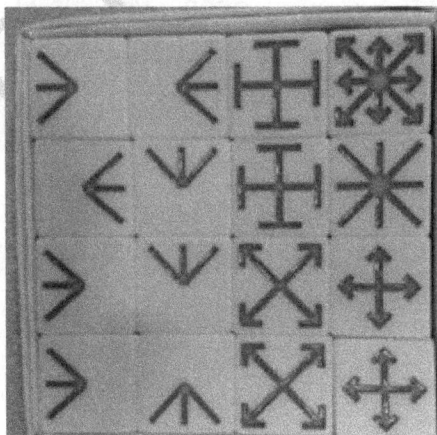

何数棋谜连锁培训及其创始人 www.homathchess.com

Frank Ho, Amanda Ho © 2015 – 2018  All rights reserved.

Student's name: _____ Assignment date: _____

## What is Ho Math Chess™ Teacher Set?

The following is an image of Ho Math Chess teaching set.

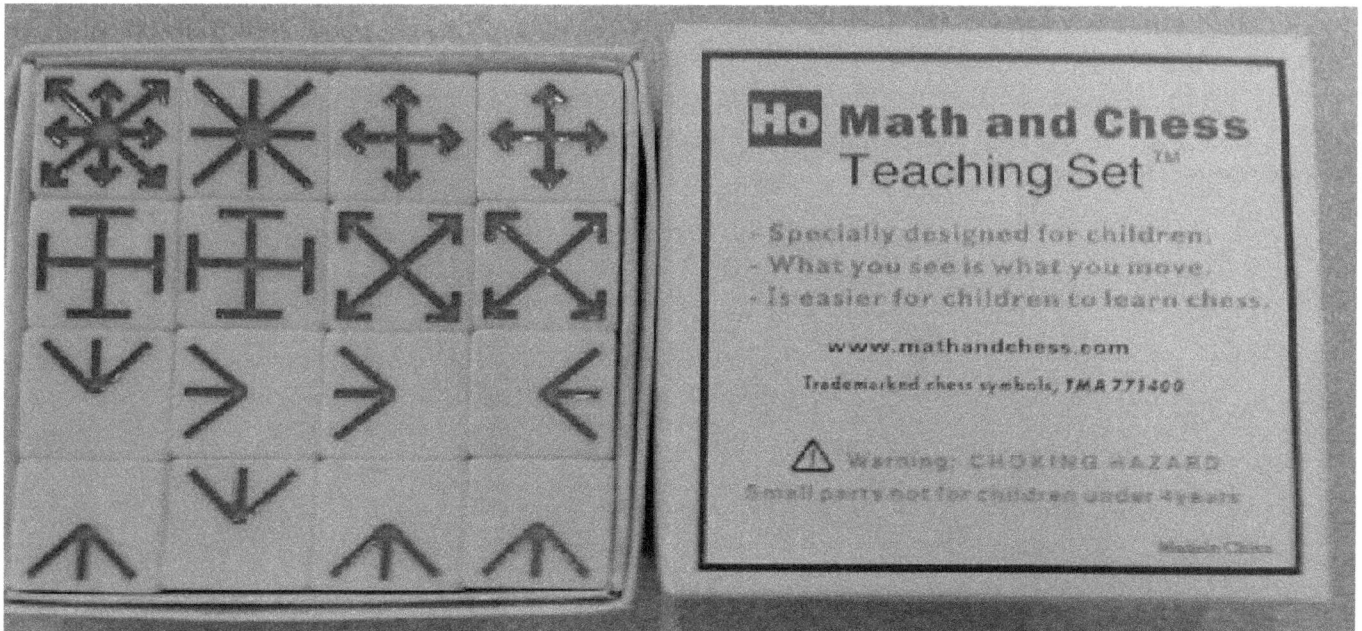

December 11, 2007, is when the world's first flat-surfaced, all-squared, uniform look chess teacher set was available to the world. This chess teacher set will become a collector's item since it is the world's first. It creates a new dimension on how math can be taught by combining chess and math through Frankho Geometry Chess Symbols (Trademark Canada TMA771400 and international copyright registered in Canada).

Student's name: _____   Assignment date: _____

**The theoretical background on how this set was published in Vector (Fall 2007, Volume 48, Issue 3), the official Journal of the British Columbia Association of Mathematics Teachers in Canada.** The website for the published article is available to view at http://www.scribd.com/doc/207579234/A-New-Chess-Set-for-Teaching-Mathematical-Chess

We use this SCL in our workbooks, and it surprisingly makes children like to work on math. The fact of the matter is it also does not cause any confusion for a student who learns math. The simple reason is they are just symbols, so they do not create any cognitive dissonance or jeopardize the learning process.

One-of-its-kind and unique Ho Math Chess Teaching Set is sold worldwide and carries a reputable brand name. This Ho Math Chess teaching set is "out-of-this-world", a genius invention using Frankho Symbolic Command Language.

何数棋谜连锁培训及其创始人 www.homathchess.com

Frank Ho, Amanda Ho © 2015 – 2018   All rights reserved.

Student's name: _____  Assignment date: _____

## Ho Math Chess Teaching Set

How to play?

Since how each chess piece could move is marked symbolically on the surface of each chess piece using the idea of symmetry of a square and the geometric concept of the line (showing by arrows) and a line segment (with no arrows), so how to move each chess piece is much clear than the "traditional" chess set.

The rook moves up/down and left/right with no number of squares restricted, so it has arrows showing it can go/down and left/right.

The bishop moves diagonally, so it has arrows showing that its chessboard only limits the number of squares that can be reached.

The same idea applies to the queen with all eight arrows in 8 directions showing the number of squares that can be reached is only limited by its chessboard, but the King has no arrows, so its movement is limited to only one square for each move.

A pawn can move forward but captures diagonally, so its moves showing no arrows (I.e., cannot move with an unlimited number of squares).

Knight has the shape of 8 L directions but no arrows, so it can "jump" to only one directed square by a line segment in L shape.

The description of our unique chess set is as follows:

- Specially designed for children as young as 4-year old to learn chess easily.
- Multi-function capabilities for playing traditional chess games and also blind or half-blind games to improve memory.
- Can be used to play puzzles, not just chess.
- Looking for something unique? Your wish has come true!
- Pocket-sized chess set for easy carrying. Children love it!
- What you see is what you move! So easy to play chess now!!!! It is a fantastic invention!
- World's first flat-surfaced international chess set with copyright-applied Geometry Chess Symbol marked on each chess piece!

Student's name: _____ Assignment date: _____

**Ho Math Chess™ Teaching Set**

An excellent chess teacher chess set

Frank Ho, the Ho Math Chess founder, invented a revolutionary chess training set specially designed and manufactured for teaching children or novice players as young as 3 or 4 years old to play chess.

This incredible chess set plays like a standard 3-D chess set but offers additional advantages. Each chess piece's moves are marked on each chess piece to make each chess piece much more manageable and fun for a young child to understand. It is a "what you see is what you move" chess set and teaches geometry concepts of lines, line segments, and transformations.

Also, since it has a flat surface, children can use "blind" chess to turn each piece upside down and then flip over one by one in a standing position to have great fun. No more spills or bumps for small hands when moving pieces.

The invention of the Ho Math Chess training set has revolutionized the chess learning population profile to possibly as young as pre-kindergartners

For more details on this innovative Ho Math Chess set, please visit www.homathchess.com.

Student's name: _____ Assignment date: _____

## How to Play Blind Chess or Half Chess

Ho Math Chess Teaching Set is specially designed for young children to learn chess. Since Ho Math Chess Teaching pieces have flat surfaces and uniform outlook in square size, the pieces cannot be identified and indistinguishable when they are turned face down.

This special feature allows children to play a special game called Blind Chess or Half Chess (or called Banqi in Chinese Chess). The rules to play Blind Chess are very similar to Chinese Blind Chess. Blind Chess is very easy and fun to play.

### Board

Blind Chess is played by two-player on half (4 by 8 square board) of the normal chessboard.

### Game Rules

The 32 pieces are shuffled, and then each of them is randomly placed face-down on each square. The first player turns over a piece, and the colour of the first piece uncovered will be the side of the first player.

### Moving a Piece

There are three kinds of moves. A player may turn a piece face-up, move a piece, or capture an opponent's piece. A player may only move face-up pieces of his or her colour. Unlike normal chess moving rules, there is one rule to move pieces in Blind Chess: a piece moves only one square up, down, left, or right. Namely, all pieces move like a rook. To capture the opponent's piece, a face-up piece may only move to a square occupied by an opponent's face-up piece.

Student's name: _____ Assignment date: _____

## Capturing an Opponent's Piece

All pieces (Black or White) are ranked according to the following hierarchy, and the capturing rule is strictly according to the defined hierarchy.

King has the highest rank and can capture the opponent's all pieces other than a pawn.
Queen can capture the opponent's all pieces other than the King.
The rook can capture the opponent's all pieces other than King or queen.
Bishop can capture the opponent's all pieces other than the opponent's King, queen, or pawn.
Knight can only capture the opponent's pawn.
Pawn has the lowest rank but can capture the opponent's King.

## How the game ends

The game ends when a player cannot make a move or until all pieces are captured. If the game is forced into an endless cycle of moves, then it is a draw.

Student's name: _____ Assignment date: _____

### How did Ho Math Chess™ get started?

Frank Ho, a Canadian math teacher, intrigued by math and chess relationships after teaching his son chess, started **Ho Math Chess™** in 1995. His long-term devotion to research has led his son to become a FIDE chess master and Frank's publications of over 20 math workbooks.

Today **Ho Math Chess™** is the world's largest and only franchised scholastic math, chess, and puzzles specialty learning centre worldwide. **Ho Math Chess™** is a leading research organization in math, chess, and puzzles integrated teaching methodology.

There are hundreds of articles already published showing chess benefits children and that math puzzles are a very good way of improving brainpower. So, by integrating chess and mathematical puzzles together to spark students` interest in math, the learning effect is more significant.

Parents send their children to **Ho Math Chess™** because they like Ho Math Chess™ teaching philosophy – offering children problem-solving questions in various formats. The questions could be pure chess, chess puzzles or mathematical puzzles in the nature of logic, pattern, three structures, Venn diagram, probability and many more math concepts.

**Ho Math Chess** has developed a series of unique and high-quality workbooks. Its first product is the world's first **Magic Chess and Fun Math Puzzles**. The workbook is not only for learning chess but also for enriching math ability. Right from the creation time, **Ho Math Chess** has set itself apart from other math learning centres, chess clubs or chess classes.

The purposes of **Ho Math Chess™** teaching method and workbooks are to:

- improve math marks.
- develop problem-solving and critical thinking skills.
- improve strategic thinking ability.
- boost brainpower.

Testimonials, sample worksheets, and franchise information can be found at www.homathchess.com.

Below is a brief of Ho Math Chess Learning Centre's history.

1995 - **Frank Ho founded Ho Math Chess Learning Centre**. It was located at #5, 5729 West Boulevard, Vancouver, with one classroom only.

## 1996

- **Ho Math Chess** moved to a bigger location. It was located at #12, 5729 West Boulevard, with three classrooms.
- **Ho Math Chess** program started to offer at St. George's Summer School in Vancouver.

## 1997

- **Mathematical Chess Puzzles for Junior** officially published.

1998 - **Fun Math Puzzles for Juniors** officially published.

1999 - **Ho Math Chess** moved to a location with exposure to the 3-storefront with five classrooms. The rooms #4 and # 6 at 2265 West 41st, Vancouver, were covered with dust when discovered by Frank Ho. Frank applied for a city license and arranged an inspection, and negotiated the lease to secure the place.

## 2000, 2001

**Ho Math Chess** math team coached by Frank Ho competed in the Mathcounts and won the third in the most competitive Vancouver region. Because of performance, the **Ho Math Chess** team was not allowed to compete provincially and forever banned from Mathcounts competition using a learning centre's name. (Note: this policy had since been relaxed in 2003.)

2003 - **Math Contest Preparation** was officially published.

## 2004

- Whole Numbers Operations - addition, subtraction published.
- Whole Numbers Operations - multiplication, division published.
- Whole Numbers Problem Solving and Puzzles published.
- Fractions, Decimals, Ratio, Equations published.
- Pre-Algebra Problem Solving Strategies published.
- **Ho Math Chess** franchise business worldwide launched.
- Franchisees Richmond, Burnaby, Ecuador joined **Ho Math Chess** Learning Centre
- **Ho Math Chess** was trademarked.

## 2005

- **Magic Chess Math Puzzles** 3rd generation workbook published.
- Creation of the **Magic Chess and Math Puzzles Prime** for pre-schoolers.
- **Ho Math Chess** trademark was approved.

Student's name: _____ Assignment date: _____

**2006**

- Article on **Enriching Math Using Chess** approved for publishing.
- Peninsula **Ho Math Chess** opened in California, USA
- Palo Alto **Ho Math Chess** opened in California, USA
- Chicago, Illinois **Ho Math Chess** opened, USA
- Frank gave a presentation on Chess in School Curriculum to Surrey teachers in Canada on Pro-D day.
- Many worldwide franchisees joined Ho Math Chess this year.

**2007**

- Master Franchise in Latin America awarded.
- Ho Math Chess continues to improve products, and this year is the breakthrough year in Ho Math Chess' growth history. Many new inventions and products are produced.
- Many worldwide franchisees have joined Ho Math Chess.

**2008**

- The master franchise of all Arabic-speaking countries awarded.
- The master franchise of Singapore was awarded.
- The master franchise of Malaysia was awarded.
- Franchise in Taiwan awarded.
- Franchise in Spain awarded.
- Franchise in Peru awarded.
- Frankho Chess Mazes invented and is running as a regular column in Vancouver Chinese paper.
- IQ Chess Brainpower workbook created for children.

**2009**

- Franchise in Mexico City awarded.
- Brainpower workbook produced
- Copyright of Geometry Chess Symbol awarded in Canada.
- Trademark of Geometry Chess Symbol awarded in Canada.

**2010**

- The additional new centre of Singapore Ho Math Chess opened.
- Some Ho Math Chess workbooks start to sell worldwide in March.
- Frankho Puzzles, including the invented Geometry Chess Symbols trademark, has been approved in Canada.
- The additional new centre of Texas Ho Math Chess opened.
- Awarded franchise in Nigeria.

Student's name: _____ Assignment date: _____

## 2011

- Released the world's first Chinese Ba-Gua math for children.
- Published world's first math, chess, and Sudoku puzzles for children workbook on www.amazon.com.
- Releases of workbooks Grade 8 and Grade 9.
- Malaysia joins Ho Math Chess.
- Starts to produce Ho Math, Chess, and Puzzles worksheets for the chess club.

## 2012

- Published math workbook for preschoolers/kindergarteners on www.amazon.com.
- Filed trademark in China.
- Successfully field-tested worksheets for chess and puzzles club.
- Completion of Malaysia franchise training.
- Malaysia Ho Math Chess grand opening launched.

## 2013

- Ho Math Chess workbooks translated into Turkish.
- Ho Math Chess Chinese trademark 何数棋谜 has been approved in China.

## 2014

- Many new Ho Math Chess workbooks are published.
- Malaysia Ho Math Chess second location is open.

Student's name: _____   Assignment date: _____

## What is the mission of Ho Math Chess?

Ho Math Chess' mission is to challenge each student's potential to the fullest and excel each student's problem-solving ability to the highest by providing a unique and enriched mathematics curriculum.

Ho Math Chess is an after-school math specialty learning centre, not a specialized chess centre. Ho Math Chess has a mission to promote math learning and help children improve or advance their math knowledge beyond their school curriculum levels. Ho Math Chess uses chess and puzzles and integrates them into math workbooks to help children learn math, improve their brain power, and spark their math interests. The consequence is children will be more interested in problem-solving, become smarter, know how to play chess, and solve puzzles.

Ho Math Chess does not vigorously promote the idea of asking children to play chess. So if a child has no interest in learning chess, then it is perfectly fine because to work on Ho Math Chess workbooks only requires children to know the basic moves of chess pieces. No chess tactics or strategies are required. It takes only a few minutes for children to grasp the knowledge of chess moves and their respective values.

## How math and chess are integrated

Many articles and research papers were published about how chess could benefit math, but none of them have shown math and chess could be integrated or combined as one worksheet for elementary students to work on. These math and chess research or articles suggest chess has been taught and math in the same classroom. We, Ho Math Chess, are the only organization that used our invention and proprietary intellectual properties to have created and developed complete math, chess, and puzzles integrated curriculum for elementary students from pre-school to grade 7.

Ho Math Chess invented and created a flat-surface chess game with chess moves marked on each chess piece. These chess symbols are internationally copyrighted (Copyright number 1095601) and trademarked (Canada trademark 771400). This chess set uses lines and line segments pointing to moves; thus, it is extremely easy for a novice to learn how to play chess as young as four years old. This idea and invention of using what we call Geometry Chess Symbol (GCS) to integrate math, chess, and puzzles are world first (2).

Student's name: _____ Assignment date: _____

## Why is Ho Math Chess™ a stand-out learning centre?

Ho Math Chess creates the world's first commercially available math and chess integrated workbook using its invention, and internationally copyright protected Ho Math Chess Geometry Chess Symbols and its Ho Math Chess teaching set by using these symbols.

GCS (Geometry Chess Symbols) is used in Ho Math Chess workbooks to link chess, math, and puzzles. GCS also has created an award-winning workbook Frankho ChessDoku, Frankho ChessMaze, and many other Ho Math Chess' math, chess, and puzzles integrated workbooks.

Ho Math Chess is a math specialty after-school learning centre emphasizing its fun math teaching methodology using integrated problem solving and math IQ puzzles workbooks.

Ho Math Chess is not just selling the idea of teaching math and chess under the same roof. We are offering integrated math, chess, and puzzles printed workbooks. In brief, Ho Math Chess has the following unique intellectual properties:

- Trade name and trademark: Ho Math Chess
- The integrated teaching methodology
- Workbooks creator and copyrights holder of the following workbooks and inventions:
  - GCS
  - Integrated math, chess, and puzzles fun workbooks using GCS.
  - Ho Math Chess teaching chess set
  - Inventions of *Frankho ChessDoku*, and *Frankho ChessMaze* and many other unique workbooks such as *Math Contest Preparation*, and *Problem Solving and Math IQ Puzzles*

Student's name: _____ Assignment date: _____

**Why are Ho Math Chess workbooks unique?**

Many math educators and research papers have shown that the idea of using a game would be a fun and effective way of teaching math. Some guidelines or lesson plans are also available for teachers to conduct a few lessons. No full-blown curriculum-based chess, puzzles, and math integrated course materials ever were developed in the past until Mr. Frank Ho took action to create the math, chess, and puzzles integrated and curriculum-based workbooks. This teaching model and methodology has been supported by a publication of over 50 research papers and over 30 workbooks in the past 20 years of Ho Math Chess' teaching and research. This is why Ho Math Chess workbooks are unique and are disruptively different from the traditional workbooks.

Ho Math Chess is the world's first and the only commercially available establishment globally, with over 20 years in math, chess, and puzzles integrated and printed workbooks to teach children from pre-school to elementary school. Teachers no longer must rely on a few scattered chess lesson plans to conduct math courses; instead, math teachers can use Ho Math Chess integrated workbooks to teach children math. Ho Math Chess workbooks are one-of-its-kind, unique, and copyrighted math, chess, puzzles truly integrated workbooks (Canada copyright no. 1069744, Chess symbols Trademark TMA771400).

Ho Math Chess materials also have different levels to suit students with different backgrounds. So, gifted students can use Ho Math Chess math contest workbooks. For remedial students, they can use fast-track workbooks.

Ho Math Chess could not create these math, chess, and puzzles integrated workbooks without using its invention that is internationally copyrighted Geometry Chess Symbol. By using this invention, math and chess are linked, and students can do math using mini puzzle-like problems. Working on Ho Math Chess worksheets, students become more interested in math than working on traditional math worksheets.

Student's name: _____ Assignment date: _____

## Ho Math Chess discovered the key linking math and chess

Ho Math Chess can overcome the bottleneck on how math and chess can be linked by discovering and creating the key which links math and chess. The key is our invented Symbolic Chess Language (SCL).

We believe that our discovery of chess moves using symmetry property on a square-shaped surface is the secret key that links math and chess. When teaching children as young as kindergartners, the difficult task is it understandably takes considerably more time for such young children to be familiar with how each piece should move since the chess figures have no clear indication of how each piece relates to symmetry. Without mastering chess moves, children cannot enjoy the joy of playing chess. Often it even becomes frustrating and discouraging for some children to pursue further. Children cannot master each chess piece's moves quickly because there is no clear relation between each chess piece's moves and its corresponding chess figurine. Our chess set is the "what you see is what you move" chess set. The SCL we invented and its new chess teaching set layout is illustrated (Figure 1).

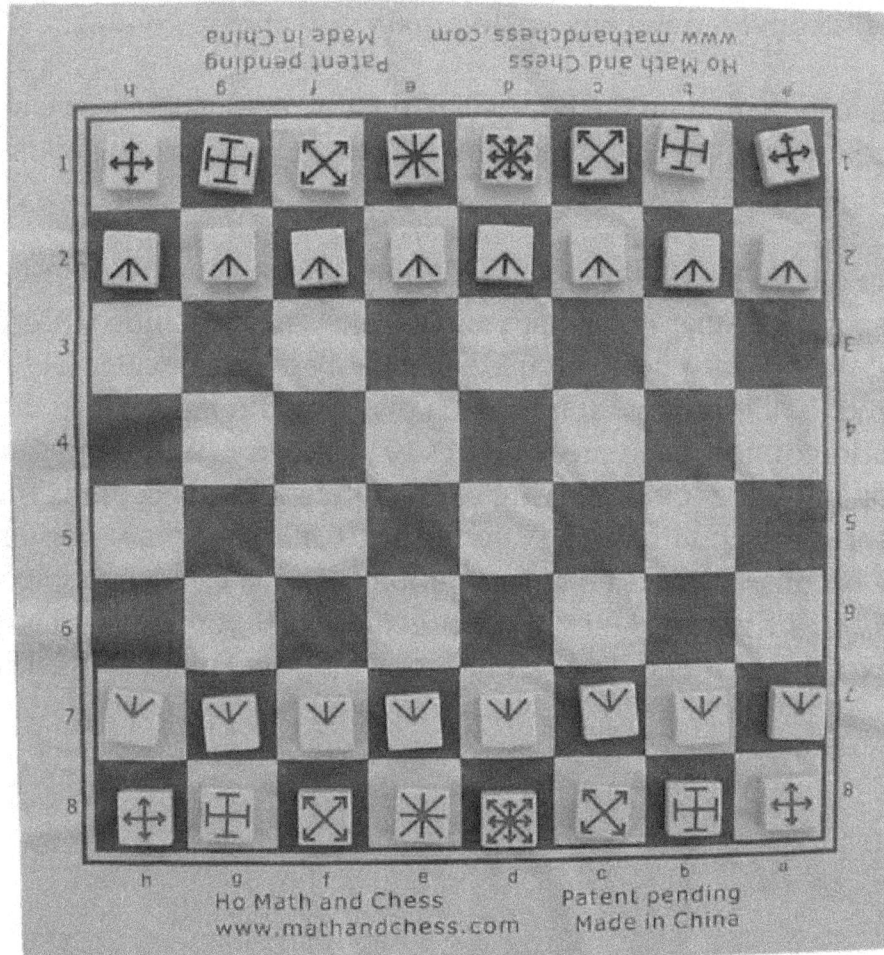

Figure 1 Geometry Chess Symbol and Ho Math and Chess Teaching Set

## 20 years of working on math, chess, and puzzles integrated workbook

We believe our innovative and unique math worksheets have revolutionized how the math workbook used to be created and has a set up a standard on how math and chess could be integrated as one worksheet. Math drill does not have to be boring. Ho Math Chess worksheets encourage students to think, so even students are doing pure computational worksheets. However, Ho Math Chess students are still encouraged to think because of our specially designed worksheets.

The following example demonstrates a reversing thinking idea to find out how many ways the number 10 can be added. The following is a copy of the above problem done by a Canadian student Meghan when she was in grade 1 in Vancouver, Canada. Meghan's answer has since been expanded in different formats in one of our workbooks, *Mom! I Learn Addition Using Math-Chess-Puzzles Connection*.

Meghan did the following sum partitioning assignment.

$$2 + 2 + 6 \qquad 0 + 10 = 10$$
$$5 + 5 \qquad 3 + 7 = 10$$
$$1 + 9 = 10 \qquad 4 + 6 = 10$$
$$\qquad\qquad 6 + 4 = 10$$
$$2 + 8 = 10 \qquad 8 + 4 = 10$$
$$\qquad\qquad 6 + 2 + 2$$
$$10 - 0 = 10$$
$$9 + 1 = 10$$
$$8 + 2 = 10$$

何数棋谜连锁培训及其创始人 www.homathchess.com

Frank Ho, Amanda Ho © 2015 – 2018  All rights reserved.

Student's name: _____ Assignment date: _____

**How Ho Math Chess worksheets integrate the Internet and IT technology into worksheets**

Ho Math Chess worksheet simulate an internet screen and internet technology

**Ho Math Chess worksheet simulates an internet screen**

Ho Math Chess worksheet layout simulates a computer screen and a cell phone screen.

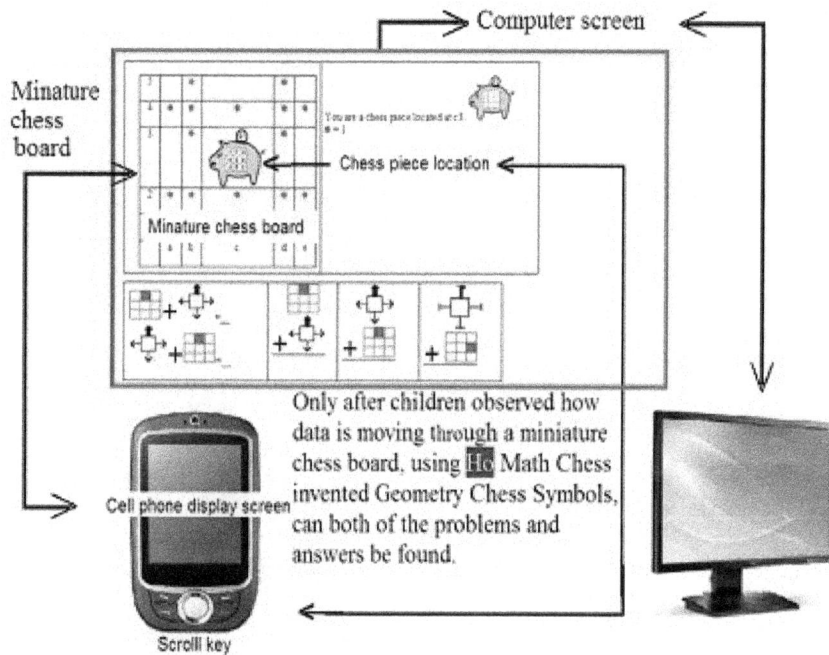

Only after children observed how data is moving through a miniature chess board, using Ho Math Chess invented Geometry Chess Symbols, can both of the problems and answers be found.

Student's name: _____   Assignment date: _____

## How Ho Math Chess worksheets integrate puzzles into worksheets

The future belongs to a generation who understands how to process information, and the information might include digits, bytes, numbers, graphics, images, languages, symbols, equations etc. How can some or most of this different nature of information processing be taught to kindergartners or primary students when learning arithmetic?

It is not an easy task; that is why there are so many different types of standalone worksheets: logic, patterns, mazes, and crosswords. These worksheets are created without interrelations with each other. This kind of isolated information processing is no longer reflecting the real world the young generation is facing today or living in the future.

The computing world children are facing today is much like a rich tapestry, where diversified fabric and colours are integrated. Children today are absorbing not just numbers but an array of information like image, sound, music, symbols, spatial information, or even abstract ideas all bundled together and delivered through many types of media. Children today are not happy just working on pure number drills without any other stimulus or motivator. Realizing the importance of having fun while learning, Ho Math Chess has been embarked on an important teaching philosophy to integrate chess and puzzles into math worksheets so that children can learn math while having fun.

When working on Ho Math Chess math worksheets, a child is acting as a data warehouse manager and sorts data through a variety of tools, namely chess, symbols, spatial relation, logic, comparison, tables, patterns, mazes, computing etc., by networking all kinds of information together. Only when children have successfully followed instructions (SCL) and, as a result, created a question themselves can a solution be found at last.

In Ho Math Chess computing worksheets, the questions are not written out for children. Still, they must be mined (after children observing how data is moving) through data warehouse (mazes), and answers must be computed by following a series of spatial relations and then analyzed using logic to reach a conclusion.

Ho Math Chess math worksheets teach children their basic computing ability and train them to be an astute data warehouse manager or excellent data miner by developing their problem-solving ability and critical thinking skills.

Student's name: _____ Assignment date: _____

## Part 2 Sample Ho Math Chess Worksheets

Here I list some of our sample worksheets that are unique and innovative.

### Category 1, Basic number facts using math-chess-puzzles connection
(何數棋謎益智健腦趣味數學)

This is to create fun arithmetic basic number facts worksheets using math-chess-puzzles combination. The following is one worksheet created in one of our workbooks.

This type of worksheets provides the training in mastering the basic number facts and cultivating each student to have a level head because it trains students to have more patience in achieving their goals.

何数棋谜连锁培训及其创始人 www.homathchess.com

Student's name: _____ Assignment date: _____

The direct teaching method is easy for an average student to follow, such as working on traditional vertical math basic numbers fact worksheets. Still, these worksheets tend to be boring and less challenging and engaging for an above-average student. The exploratory teaching method is more challenging and involves multi-step problems, so it is not easy for an average to grasp sometimes.

Our integrated worksheets take the exploratory approach, which trains an average student to use more of their brains. The problem is not difficult and teaches students a systematic and efficient calculation method to get answers instead of offering multiple methods to confuse most of the students.

The following problem taken out from this workbook demonstrates our idea as follows:

Adding up to 12

| 5 | ? | 4 | ? | ? | ? |
|---|---|---|---|---|---|
| 4 | 2 | 9 | 3 | 2 | 5 |
| 3 | ? | 6 | 2 | 4 | ? |
| 2 | ? | 8 | 5 | 7 | ? |
| 1 | ? | ? | ? | 7 | ? |
|   | a | b | c | d | e |

You are at square c3

$= \square$ .

Mini chessboard teaches coordinates c3.

$\square + \boxed{} + \boxed{} = 12$    $\square + \boxed{} + \boxed{} = 12$

2  +  3  +  7  = 12    2  +  2  +  8  = 12

## Many Ho Math, Chess, and puzzles integrated worksheets posses the following properties:

Multi-task: Students learn to lookup table to retrieve information using mini chessboard.

Multi-direction: Rook and Bishop moves

Multi-concept: Addition and subtraction for the same problem such as 2 + 3 + ? = 12.

Multi-strategy: Working forward and backwards.

Multi-operation: Addition and subtraction sone at the same time.

Multi-step: Students need to figure out what is the question before working on the solution by following abstract symbols.

Multi-sensory: Chess provides hands-on experience and trains eyes, hands, and brain coordination. Chess is part of math curriculum

Student's name: _____ Assignment date: _____

**Category 2, Integrated Math, Chess, and Sudoku** (何數棋謎趣味數學)

This math, chess, and puzzles integrated worksheets have a mini chessboard on the top left and a Sudoku puzzle on the top right. The math computation problems do not even exist until the student has worked out the puzzle. This is one of my favourites. Initially, when I first created a prototype of this type. I only had one number located at e2, so it tends to be repetitious. Later I solved the problem by creating two matrices to replace the single number.

何数棋谜连锁培训及其创始人 www.homathchess.com

Frank Ho, Amanda Ho © 2015 – 2018  All rights reserved.

Student's name: _____ Assignment date: _____

You are at (e, 2) = ⊞

| 3 (a, 3) | (b, 3) | (c, 3) |
|---|---|---|
| 2 (a, 2) | **4 1** / **2 3** | (c, 2) |
| 1 (a, 1) | (b, 1) | (c, 1) |
| d | e | f |

Rule: All the digits 1 to 3 must appear exactly once in every row and column.

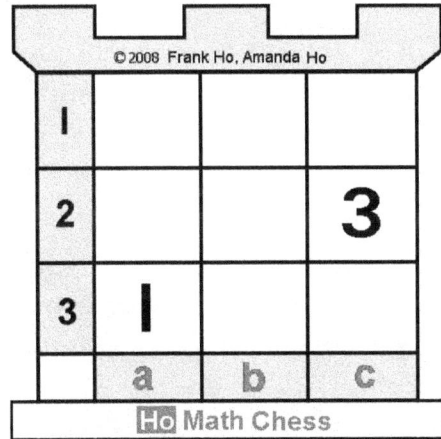

© 2008  Frank Ho, Amanda Ho

|   | a | b | c |
|---|---|---|---|
| 1 |   |   |   |
| 2 |   |   | 3 |
| 3 | 1 |   |   |

Ho Math Chess

Student's name: _____ Assignment date: _____

### Category 3, Frankho ChessDoku

Integrated Math, Chess, and Sudoku

Frankho ChessDoku is solved using one or more operators of addition, subtraction, multiplication, or division after following chess moves and logic.

Rule

All the digits 1 to 3 must appear exactly once in every row and column. The number that appears in the bottom right-hand corner is the result calculated according to the arithmetic operator(s) and chess move(s) as indicated by the darker arrow(s).

Student's name: _____ Assignment date: _____

## What is Frankho ChessDoku?

The following is a sample of Frankho ChessDoku.

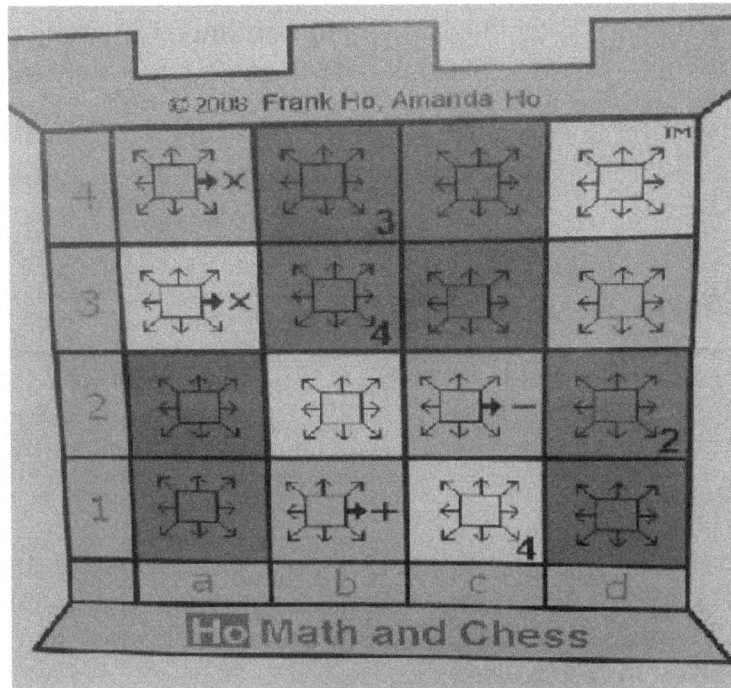

*Frankho ChessDoku* was invented by a Canadian math teacher Frank Ho (1, 2). Seeing the popularity of Sudoku but with no computation capability, Frank decided to do something about it, so Frank used his invented Geometry Chess Symbols (Canada Trademark TMA771400, copyright 1069744) along with Sudoku created the *Frankho ChessDoku* in 2008. *Frankho ChessDoku* is a unique puzzle that combines chess and Sudoku and is specially designed for children to solve arithmetic Sudoku by following chess moves (1). In 2009 Frank Ho and his wife Amanda Ho published a *FrankHo Math Chess Puzzles for Children* workbook. This workbook is now available from www.amazon.com.

Frank always has an idea about teaching math: students should always be encouraged and given a chance to THINK, and it means even when they are doing pure computation problems. This is why he has created many computation workbooks without obvious problems presented to children; instead, children must figure out what to do by going through a puzzle-like thinking process. *Chess + Sudoku = Fun Frankho ChessDoku*

The pleasure of working on these kinds of workbooks could be very well described by a famous classical Chinese poem 山重水复疑无路,柳暗花明又一村(Equivalent English phrase is *seeing the light at the end of the tunnel.*)

Frank has described the feeling, in Chinese rhyming sentences (打油诗), when working on our math, chess, and puzzles-integrated workbooks. Its meaning is mainly to describe the miracle of puzzles.

只见棋谜不见题　劝君迷路不哭涕
数学象棋加谜题　健脑思维真神奇

**Introduction of CalcuDoku**

The original CalcuDoku was invented in 2004 by a Japanese teacher Tetsuya Miyamoto in Japan (3).

**Comparisons**

The critical difference between *Frankho ChessDoku* and CalcuDoku is that *Frankho ChessDoku* uses Frank's invented Geometry Chess Symbols to guide children on the directions of arithmetic operations instead of using "boxes or "cages" as used in Miyamoto's puzzles (now called kenKen in North America).

*Frankho ChessDoku* does not just use chess pieces to replace numbers in Sudoku, as seen in some ChessDoku puzzles. *Frankho ChessDoku* invites children to trace chess moves to see the results just as if they were playing a chess game by examining the intersections of chess moves and then use the logic of Sudoku to figure out the answers. Both strategies of playing a chess game, especially the intersections of chess moves, and the arithmetic Sudoku logic need to be combined to solve *Frankho ChessDoku* puzzles.

Miyamoto runs a learning centre in Japan and teaches his puzzles to children. Frank and his wife also use their puzzles to teach children from age four and up in their learning centre in Vancouver, Canada. Both Frank and his wife teach kindergarten to grade 12 math, and both also teach math contest preparations.

From a student's learning math point of view, *Frankho ChessDoku* offers more powerful learning and mental training advantages over regular Sudoku and other types of arithmetic Sudoku. The following table gives some comparisons. In addition to being a fun puzzle, *Frankho ChessDoku* is more suitable for improving their brainpower and mental math ability.

何数棋谜连锁培训及其创始人 www.homathchess.com

Frank Ho, Amanda Ho © 2015 – 2018   All rights reserved.

Student's name: _____  Assignment date: _____

|  | *Frankho ChessDoku* | Regular Sudoku | CalcuDoku |
|---|---|---|---|
| Plus, minus, multiplication, and division | Can provide four mixed operations by following chess moves within one equation with no confusion. | No computations | Only independent and separate +, −, ×, ÷ operations can be provided. Mixed four basic operations could confuse young children. |
| Vertical, horizontal, and diagonal operations | The horizontal or vertical operations are provided. The diagonal operations can be provided. The "jump" operation (knight move) can be provided. | No computations | Only horizontal or vertical operations are provided. No diagonal operations can be provided. No "jump" operation can be provided. |
| Framed or boxed operations | No framed operation is required since chess moves guide the operation direction. Children can always circle operation statements themselves. This flexibility allows intersecting "boxes" with no confusion. | No computations | Since the operation is always "boxed" or "caged" with a single operation, no flexibility is allowed for mixed operations. Intersecting frames or boxes would cause confusion. |

Student's name: _____  Assignment date: _____

## Example 1

### Rule

All the digits 1 to 3 must appear exactly once in every row and column. The number that appears in the bottom right-hand corner is the result calculated according to the arithmetic operator(s) and chess move(s) as indicated by the darker arrow(s).

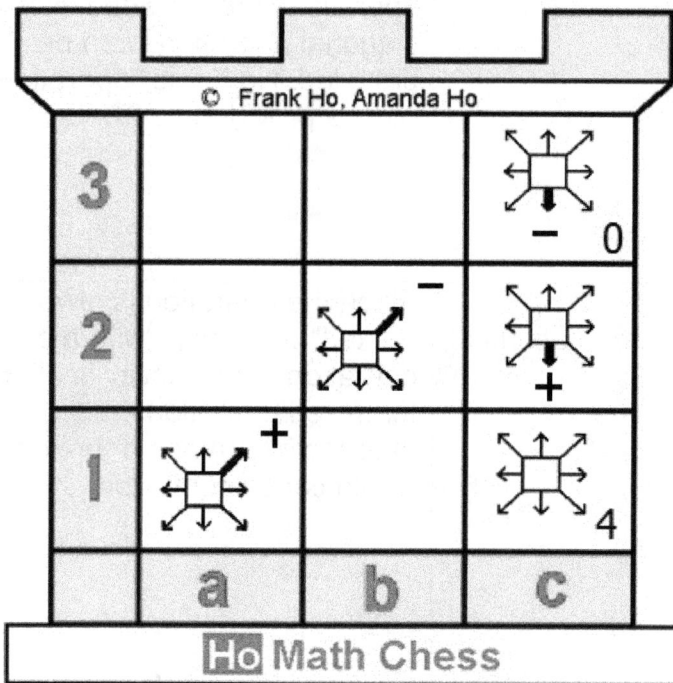

### Step 1:

Circle all operations by following chess moves.

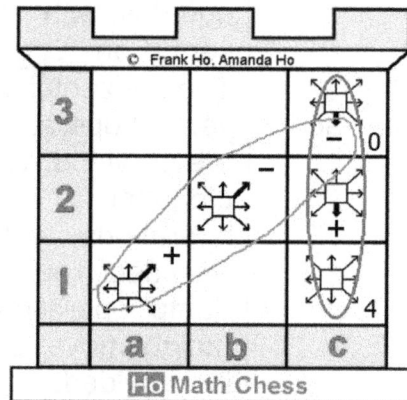

### Step 2.

For the diagonal oval, $2 + 1 - 3 = 0$, $1 + 2 - 3 = 0$.

For the vertical oval, $3 - 1 + 2 = 4$
So, we know c3=3.

The final answer is as follows.

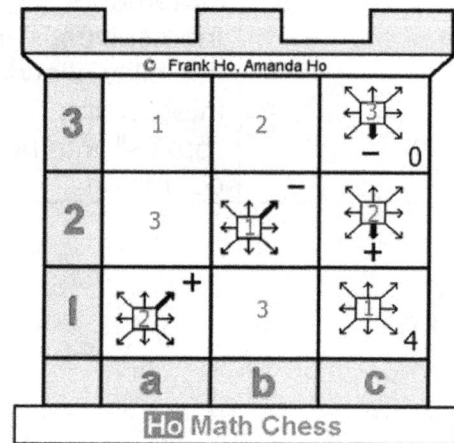

Student's name: _____ Assignment date: _____

A CalcuDoku is not able to produce the diagonal operation of the above *Frankho ChessDoku*. The mixed operation $3 + 2 - 1 = 4$ or $2 + 3 - 1 = 4$ is also difficult for children to work on if it happens in the CalcuDoku since it involves two operations simultaneously but is very easy for *Frankho ChessDoku* to identify it with no confusion. This deficiency in CalcuDoku means children seem to be stuck with only one operation at a time with very little chance to work on mixed operations. On the other hand, Children working on *Frankho ChessDoku* will have plenty of opportunities to work on either single or mixed operations with no confusion only by following chess moves.

Student's name: _____ Assignment date: _____

**Example 2**

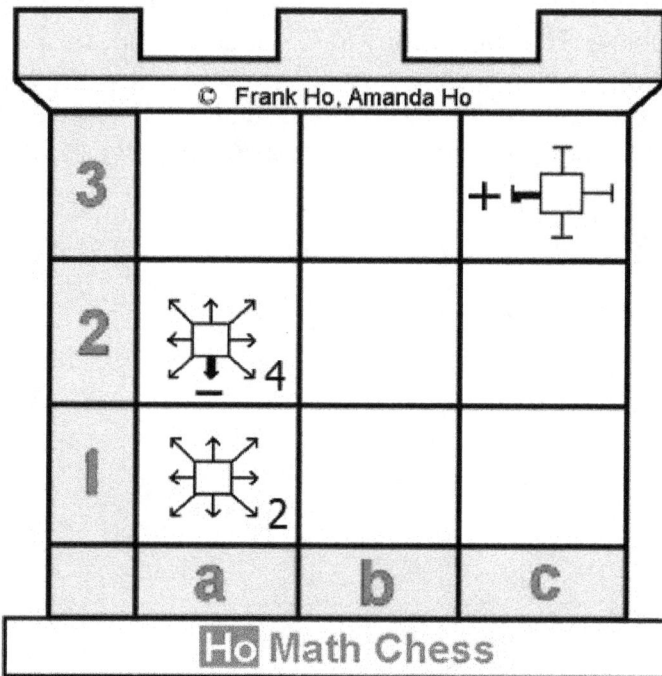

**Step 1**

Circle all operations.

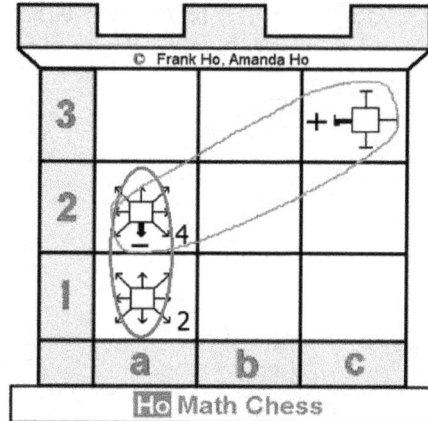

**Step 2**

For the diagonal oval, 1 + 3 = 4, 3+1 = 4, 2 + 2 = 4.

For the vertical oval, 3 − 1 = 2.
So, we know a2 = 3.

The final answer is

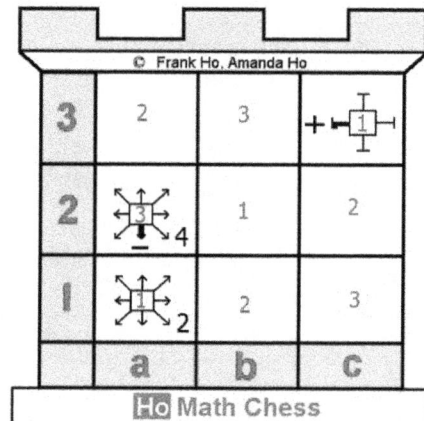

A CalcuDoku cannot produce the above "Jump" movement as acting by the chess knight move at c3.

Student's name: _____ Assignment date: _____

## Example 3

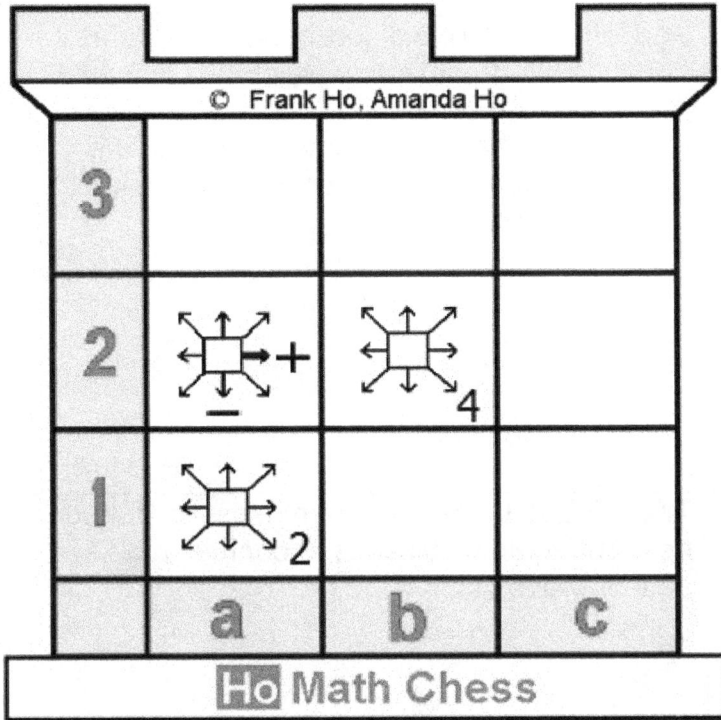

Step 1

Circle all operations.

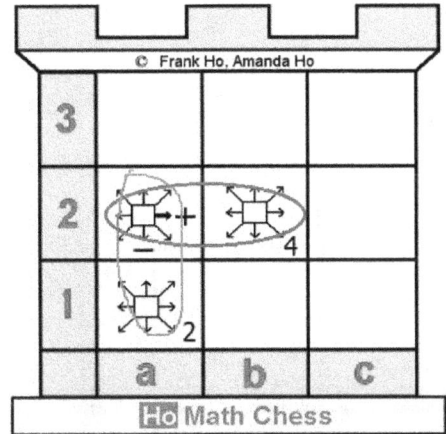

Step 2

Start at intersection a2.
For the horizontal oval, 3 + 1 = 4, 1+ 3 = 4, 2 + 2 = 4.
For the vertical oval, 3 − 1 = 2
So, we know a2 = 3.

The final answer is as follows:

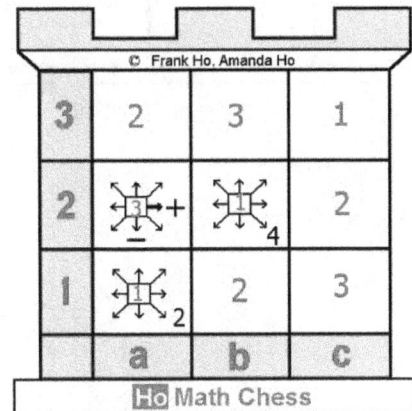

This world-famous Sum and Difference problem can be easily illustrated using the above *Frankho ChessDoku* diagram with the intersection. It can be very easily solved but trying to create it using the idea of CalcuDoku demonstrates confusion for children.

Frank Ho, Amanda Ho © 2015 – 2018  All rights reserved.

Student's name: _____ Assignment date: _____

The following are the same problem (Sum and Difference) using the diagrams of CalcuDoku.

| | |
|---|---|
| | The left CalcuDoku diagram causes confusion because we do not know which box is for 4 + and which box is for 1−. The Venn diagram concept can be easily demonstrated in the *Frankho ChessDoku* but causes confusion in the CalcuDoku. |
| | The left CalcuDoku diagram uses the dotted box, but again it still causes confusion as stated above. |

Student's name: _____ Assignment date: _____

## Commutative law

The conventional way of calculating is in the direction of left to right or top to down. Still, this rule does not apply to CalcuDoku because, as shown below, 2 − can be expressed as 3, 1 or 1, 3, and it appears to students that the subtraction can be done by exchanging the two numbers, and this is in violation of the commutative law. It would have no problem for *Frankho ChessDoku* to handle the subtraction and division operators because the calculation direction is clearly defined by using chess moves.

The answer could be operated from left to right for the subtraction operator, but sometimes, it could also be from right to left.

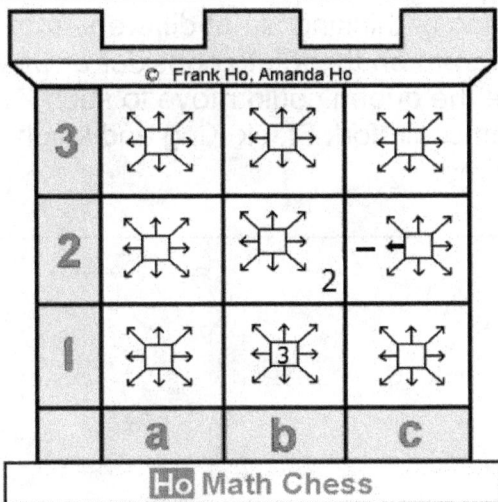

The above CalcuDoku requires the student to think about how 3 and 1 are to be arranged to present extra challenges. Still, the confusion could also occur when the mixed operators (+, −, ×, ÷) are presented together with no operating directions are given.

The left *Frankho ChessDoku* presents no operation confusion and does not have the confusion of commutative law for children.

Student's name: _____ Assignment date: _____

## Chess strategy and *Frankho ChessDoku* strategy

Often a chess player would analyze the chess moves and see where each chess piece intersects with each other, then decide to take the action of the next move. This kind of thinking is also reflected in the strategy on how to solve *Frankho ChessDoku,* and the following example demonstrated the transferred knowledge between chess and *Frankho ChessDoku.*

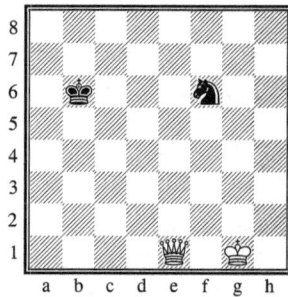

Find a Black move to fork.
Qe1 moves to f2 to fork black king and knight.

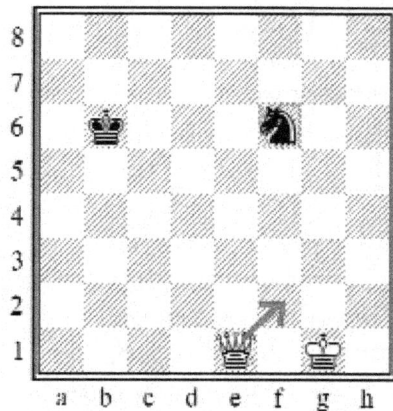

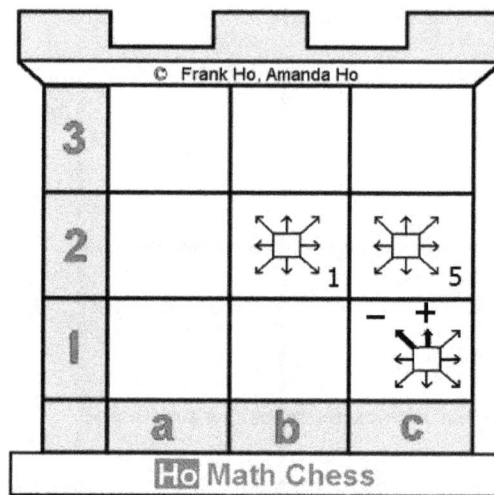

The above c1 intersects with b2 and c2 simultaneously, so in other words, c1 is a square where bishop and rook intersect in chess. This kind of thinking is not different from the chess diagram on the left to consider at what square where the queen could move to such that the queen could fork Black King and knight at the same time.

Student's name: _____ Assignment date: _____

## Triangular solving strategy for three by 3 grid

The simplest 3-by-3 case of *Frankho ChessDoku* can be created using only one number and one math operator. All other math operations are redundant.

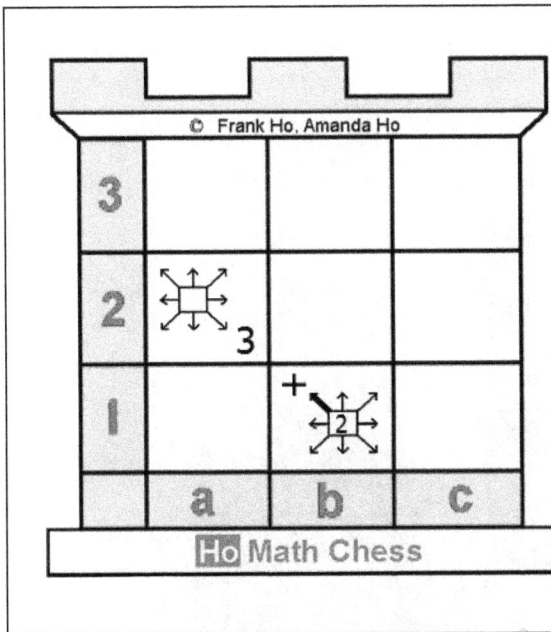

The triangular method can be used to decide the number at b2, which must be 3.

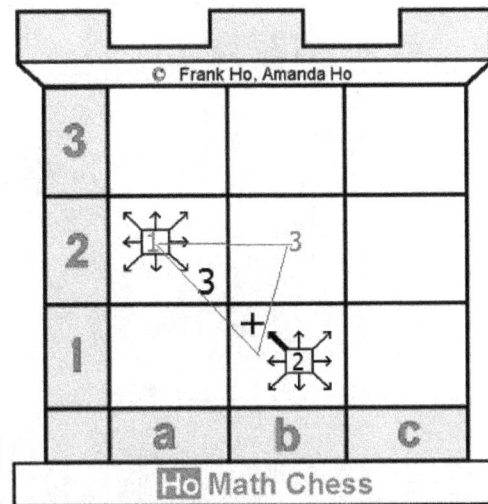

CalcuDoku cannot duplicate the above diagonal operation, but a vertical and horizontal operation can be made.

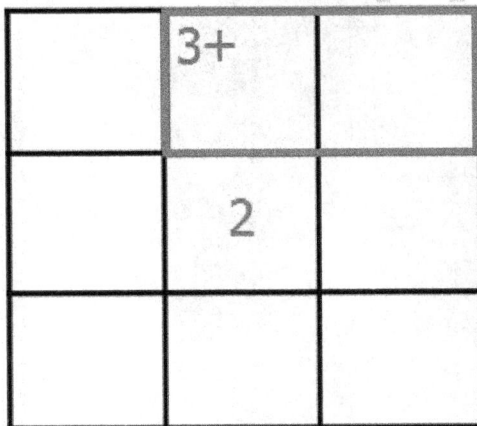

*Frankho ChessDoku*

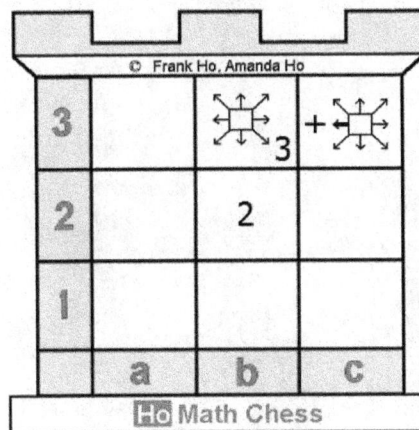

何数棋谜连锁培训及其创始人 www.homathchess.com

Frank Ho, Amanda Ho © 2015 – 2018   All rights reserved.

Student's name: _____ Assignment date: _____

## References

(1) http://susanpolgar.blogspot.ca/2008/10/ho-math-and-chess.html
(2) http://www.mathandchess.com/releases/release/1441781/17465.htm
(3) http://en.wikipedia.org/wiki/Tetsuya_Miyamoto

Students can frame their Frankho ChessDoku as follows.

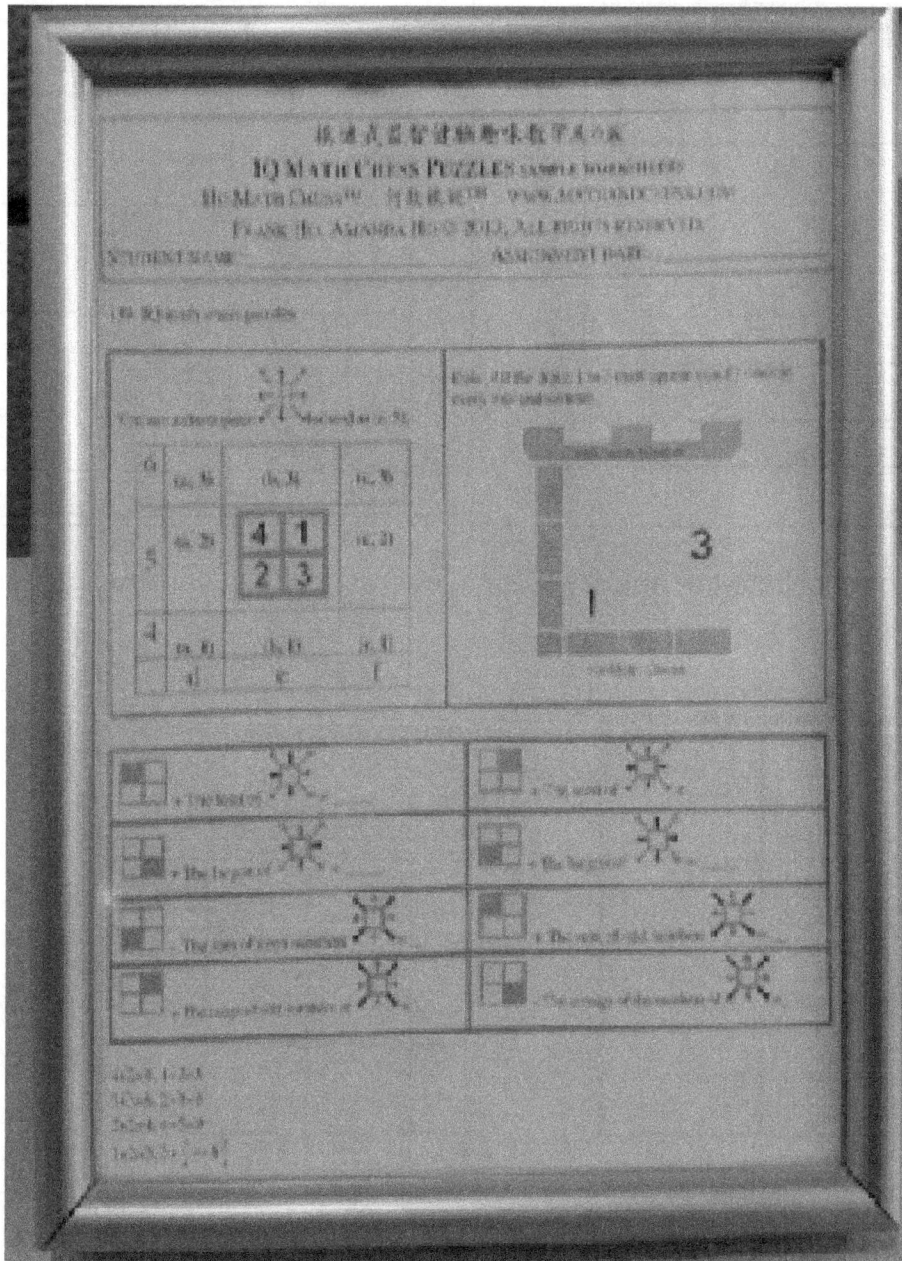

Student's name: _____ Assignment date: _____

**Category 4, Frankho ChessMaze** 何數棋謎宮

- **Trace the path from ⊠ to ♚.**
- **A darker line segment shows movement direction.**

| Transformation | Symbol | Example |
|---|---|---|
| **Slide** | $[\text{-} \leftrightarrow +, \updownarrow]$ | [3, -2] **Move right 3 squares and down 2 squartes.** <br> [0, 2] **Move up 2 squares.** |
| **Rotation** | ⌐ ⌐ | ⊕ = ⊕, ⊕ = ⊕ |
| **Flip** | ⊕ flip | ⊕ flip = ⊕, ⊕ flip = ⊕ |

**PW (password): Calculate 1 + 2 + 3 + 4 + … + 98 + 99 + 100 = _____**

⊠ = ♛   ⊕ = ♜   ⊠ = ♝   ⊥ = ♞

Student's name: _____ Assignment date: _____

## Category 5, Incorporating abacus into math, chess, and puzzles

Abacus is integrated into Ho Math Chess Worksheets

You are a chess piece at c3. ● = 1

Student's name: _____ Assignment date: _____

## Category 6, Frankho Cube Math

How Ho Math Chess worksheet integrates virtual toy (3d Cube) into math worksheets

(Step 1, Solve Sudoku. Step 2. Fold-down a cube and solve a puzzle)

Folded down back view
Write a letter on each face, which is corresponding to the cube's net.

Folded down front view

Rule: Each row, column, and cell must have only one digit from 1 to 9.

| 9 | 2 |   | 5 |   |   | 7 |   |   | 6 |
| 8 | 4 |   |   | 9 | 6 |   |   | 2 |   |
| 7 |   |   |   | 8 |   |   |   | 4 | 5 |
| 6 | 9 | 8 |   |   | 7 | 4 |   |   |   |
| 5 | 5 | 7 |   | 8 |   | 2 |   | 6 | 9 |
| 4 |   |   |   | 6 | 3 |   |   | 5 | 7 |
| 3 | 7 | 5 |   |   | 2 |   |   |   |   |
| 2 |   | 6 |   |   | 5 | 1 |   |   | 2 |
| 1 | 3 |   |   | 4 |   |   | 5 |   | 8 |
|   | a | b | c | d | e | f | g | h | i |

+ _ = _ + _ = _

+ _ = _ + _ = _

+ _ = _ + _ = _

+ _ = _ + _ = _

+ _ = _ + _ = _

+ _ = _ + _ = _

+ _ = _ + _ = _

+ _ = _ + _ = _

Student's name: _____ Assignment date: _____

Frankho Cube Math™

| Large cube net | It folded down the front view of the large cube. | face | symbol | value | | face | symbol | value |
|---|---|---|---|---|---|---|---|---|
| | | A | □ 7 | $\frac{1}{4}$ of 28 | | D | ■ 4 | $\frac{1}{5}$ of 20 |
| | | B | △ 4 | $\frac{1}{2}$ of 8 | | E | △ 6 | $\frac{1}{6}$ of 36 |
| | | C | ○ 9 | $\frac{1}{3}$ of 27 | | F | ● 8 | $\frac{1}{7}$ of 56 |

| Small cube net | It folded down the front view of the small cube. The small cube is inside the large cube. | face | symbol | | face | symbol |
|---|---|---|---|---|---|---|
| | | A | ■ +1  5 | | D | △ −3  3 |
| | | B | 20 − ○  11 | | E | 10 − ●  2 |
| | | C | □ × 4  28 | | F | △ ÷ 3  $\frac{4}{3}$ |

The face value is decided by symbol, and the small cube face value is obtained from the large cube face with the same symbol.

| | |
|---|---|
| + = ___ + ___ = 5+9=14 | + = ___ + ___ = ___ |
| + = ___ + ___ = 4+3=7 | + = ___ + ___ = ___ |
| + = ___ + ___ = 28+4=32 | + = ___ + ___ = ___ |
| + = ___ + ___ = 2+7=9 | + = ___ + ___ = ___ |

何数棋谜连锁培训及其创始人 www.homathchess.com

Student's name: _____ Assignment date: _____

## Category 7, Amandaho Moving Dots Puzzle

**Amandaho Moving Dots Puzzle ™ (移点子)**

You are a rook at c3.
Move some dots in c4, c2, or d3 squares into c3 square such that the sum of dots + dots in each of rook's moves at c3 will be equal to the number shown on its destination square. See the following example.

Example

Problem

Student's name: _____ Assignment date: _____

## Category 8, Multi-grade multi-level math (多年级多功能计算题)

How do you create one page of calculation worksheet which can be used for multi-grade students from grade 1 to grade 7? The following is what we have created for across grades teaching in the same room at the same time.

只见棋谜不见题   劝君迷路不哭涕   数学象棋加谜题   健脑思维真神奇

You are a multi-faced chess piece located at c3.

| | | ▣ | ▭ | ⊙ | △ |
|---|---|---|---|---|---|
| Fraction | $2\frac{2}{5}$ | 24 | $3\frac{2}{3}$ | $4\frac{2}{3}$ | $5\frac{3}{4}$ |
| decimal | 0.0024 | 0.024 | 10000 | 0.0048 | 0.001 |
| Whole | 5 | 15 | 20 | 10 | 25 |
| % | 100 | 25% | 10% | $33\frac{1}{3}$% | 0.05% |

**Whole number**

**Decimal**

**The fraction of addition and subtraction [Must have the same measuring unit (denominator).]**

**The fraction of multiplication and division [Do not need to have the same measuring unit (denominator).]**

**%**

何数棋谜连锁培训及其创始人 www.homathchess.com

Frank Ho, Amanda Ho © 2015 – 2018   All rights reserved.

Student's name: _____   Assignment date: _____

## Category 9, the robot is integrated with math

Control panel: you are located at c3.

| 5 | | | | |
|---|---|---|---|---|
| 4 | 🤖 | 🔲 | | 🤖 |
| 3 | 🤖 | | | 🔲 |
| 2 | 🤖 | | 🔲 | 🤖 |
| 1 | a | b | c | d | e |

| | | | |
|---|---|---|---|
| Step 1 | ↕↔ | Step 5 | ⤢ |
| Step 2 | ↔ | Step 6 | ↔ |
| Step 3 | ↔ | Step 7 | ⤢ |
| Step 4 | ⤢ | Step 8 | ⤢ |

**Antenna Magic Square***

Place a number in each square box so that each row, column, and main diagonal has the same sum.

| | |
|---|---|
| b2 | Replace ? by a number on the arm. |
| b3 | Replace ? a number on the leg and pelvis. |
| b4 | Find out how many O's to replace ? in Cindy's unbalanced scale tray. |
| c2 | Complete robot Cindy's antenna magic square* for numbers of 1, 3, 5, 7, …, 15, and 17. |
| c4 | Complete robot Andy's antenna magic square* from 1 to 9. |
| d2 | Replace each letter with a digit on all robot chests. The same letter means the same digit for the same robot. |
| d3 | Complete robot Bob's antenna magic square* from 0 to 8. |
| d4 | Replace ? by many antennae |

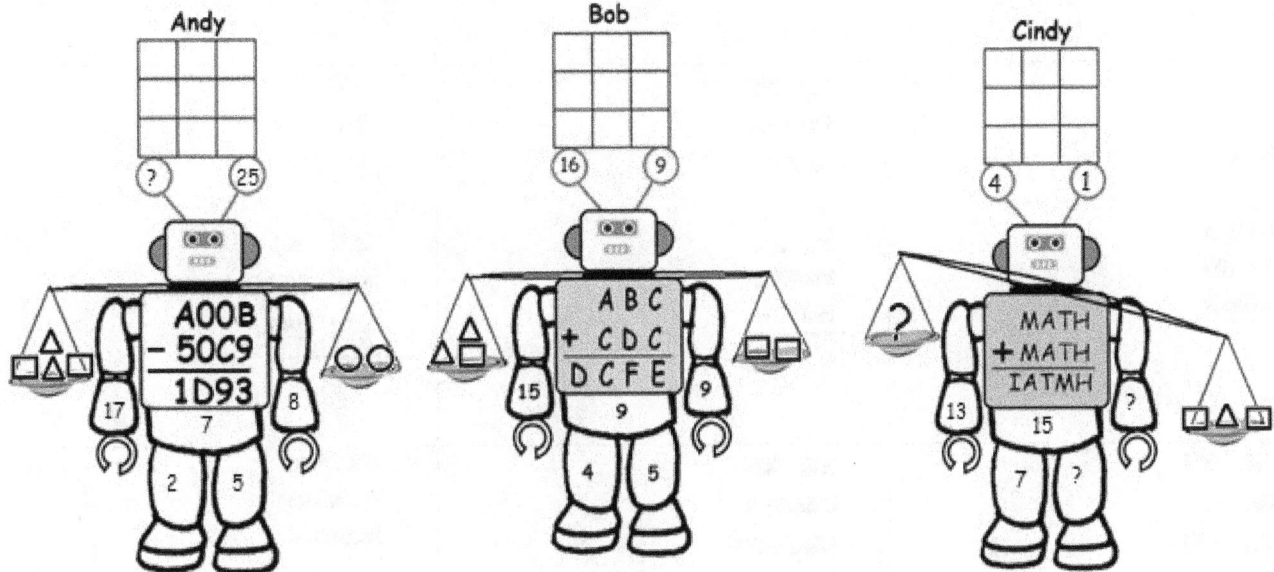

Andy

| | | |
|---|---|---|
| | | |
| | | |
| | | |

? | 25

A0OB
− 5OC9
1D93

17 | 8
7
2 | 5

Bob

| | | |
|---|---|---|
| | | |
| | | |
| | | |

16 | 9

A B C
+ C D C
D C F E

15 | 9
9
4 | 5

Cindy

| | | |
|---|---|---|
| | | |
| | | |
| | | |

4 | 1

MATH
+ MATH
IATMH

? | ?
13 | 15
7 | ?

Student's name: _____ Assignment date: _____

**Ho Math Chess also incorporates Chinese Ba-Gua and Chinese Tangram into math worksheets.**

No Part of this publication can be copied, duplicated, or reproduced.

52

何数棋谜连锁培训及其创始人 www.homathchess.com

Student's name: _____ Assignment date: _____

## Category 10, **Frankho Eight Diagrams Math™**

何氏八卦数学™ (The Frankho Ba Gua Math™)

七巧板八卦九宫数　不见题只现谜中谜　何氏国际象棋谜踪　金球第一独特发明

见棋谜不见题　劝君迷路不哭涕　数学象棋加谜题　健脑思维真神奇

You are positioned in the middle of a chessboard, which is at c3, and it is called Ba-Gua (八卦) in Chinese. "Ba" means eight (8). "Gua" means trigram. The Ba-Gua is divided into 8 Gua (trigrams/sections) consisting of the most original and basic binary form of "Yin" (female - two short hyphens) and/or "Yang" (male - one long hyphen). The following octagon is Ba-Gua, which will guide you to solve the mysterious mathematical chess puzzle. First, you solve mathematical problems according to the eight symbols (made of broken and unbroken lines) of Ba-Gua pointing to. You get each answer by associating it with the Yin Yang (Taiji) in the middle of Ba-Gua. Finally, you will go back to the chessboard, which has Ba-Gua in the middle as a controller, and it shows how you shall move by following the chess symbols to get the final answer.

This is Ba-Gua (八卦).

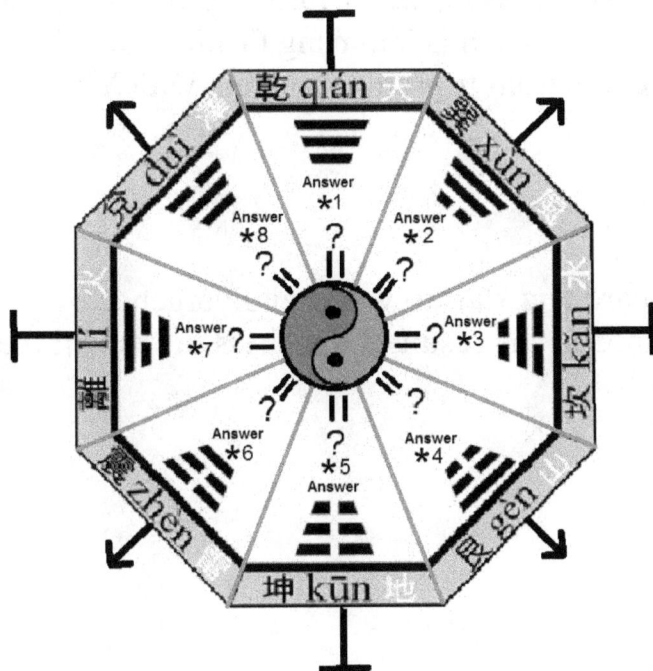

| 000 Earth | 100 Mountain | 010 Water | 110 Wind |
|---|---|---|---|
| 坤 | 艮 | 坎 | 巽 |

| 震 | 離 | 兌 | 乾 |
|---|---|---|---|
| Thunder 001 | Fire 101 | Lake 011 | Heaven 111 |

There is a close relationship between Ba-Gua and chessboard. A queen can move eight directions, which are the same as the eight trigrams on the Ba-Gua. The 64 squares of a chessboard are the same as the number of 64 hexagrams of the Chinese I Ching.

Because of Ba-Gua and chess's close relationships, Ho Math and Chess has created a world-first mathematics puzzle that links Ba-Gua, chess, and mathematics.

Step 1

Students have a condensed version of the chessboard (5 by 5), and in the middle of the chessboard has a diagram of Ba-Gua. Each trigram guides students on which math problem shall be solved. The math problem consists of 8 major areas: Chinese Tangram, Frankho Puzzle, Chinese magic square, smart and speedy computation, word problems, Olympia math contest problems, Math IQ problems, and recreational math problems.

Once the problems are solved, they must pass back to Ba Qua for further processing. Each answer is passed on to the original of the universe, which is YinYang (Taiji). An intermediate result is obtained by using its answer from trigram to interact with Yin Yang's answer.

Step 2

At this point, you shall go back to the chessboard and start to work on divergence by following the chess move. Each answer obtained as the result of the chess move will then be processed with step 1 to get the final answer.

Student's name: _____ Assignment date: _____

You are at c3.

| | | | | | |
|---|---|---|---|---|---|
| 5 | | 15 | | 12 | |
| 4 | 10 | 19 | | 11 | 13 |
| 3 | | | | | |
| 2 | 15 | 27 | | 23 | 14 |
| 1 | | 16 | | 11 | |
| | a | b | c | d | e |

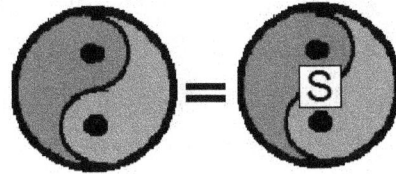

S= 64

| *1= + | *2 = + | *3 = + | * 4 = + |
|---|---|---|---|
| *5= + | *6 = + | *7 = + | * 8 = + |

Student's name: _____ Assignment date: _____

Frankho Eight Diagrams Math™　何氏八卦数学™ (The Frankho Ba-Gua Math™)

七巧板八卦九宫格　不见题只现谜中谜　何氏国际象棋谜踪　金球第一独特发明

见棋谜不见题　劝君迷路不哭涕　数学象棋加谜题　健脑思维真神奇

You are at c3. Work from inner convergence to outer divergence. 从百川入海到一门独秀

Answer _____

Answer _____

Answer _____

Answer _____

何数棋谜连锁培训及其创始人 www.homathchess.com

Frank Ho, Amanda Ho © 2015 – 2018   All rights reserved.

Student's name: _____ Assignment date: _____

Frankho Eight Diagrams Math™　何氏八卦数学™ (The Frankho Ba-Gua Math™)

You are at c3. Work from inner convergence to outer divergence. 从百川入海到一门独秀 You are at c3.

Work from inner convergence to outer divergence. 从百川入海到一门独秀

Answer _____

Answer _____

Answer _____

Answer _____

**☰111**

Adam has 101 apples, and Bob has 15 apples. How many apples does Adam have to give Bob, so they have an equal number of apples?

Answer _____   Return the answer to Ba-Gua.

| Algebraic method. | Arithmetic method |
|---|---|
| | |

**☷110**

| You are at b2, and the area of the entire square | The Chinese Tangram. (七巧板; literally "seven boards of skill") invented by the ancient Chinese hundreds of years ago is a dissection puzzle consisting of seven flat shapes (2 large right triangles, 1 medium right triangle, 2 small right triangles, 1 square, and 1 parallelogram) which are put together to form shapes. The puzzle's objective is to form a specific shape (given only an outline or silhouette) using all seven pieces, which may not overlap. |
|---|---|

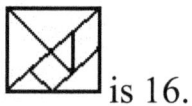

 is 16.

| 3 | 9 | 13 | 15 |
|---|---|---|---|
| 2 | 18 |  | 12 |
| 1 | 17 | 16 | 14 |
| | a | b | c |

Return the last answer to Ba-Gua.

何数棋谜连锁培训及其创始人 www.homathchess.com

Frank Ho, Amanda Ho © 2015 – 2018   All rights reserved.

Student's name: _____ Assignment date: _____

---

**☰010**

Frankho ChessDoku

All the digits 1 to 3 must appear in every row and column. The number that appears in the bottom right-hand corner is the result calculated according to the operator(s) and chess move(s).

Return the answer ⟡₁ to Ba-Gua.

© 2008 Frank Ho, Amanda Ho

---

**100**

Fill each box with a number. The same shape box has the same number.

$$29 - \square - \triangle = 21$$

$$\square = \triangle + \triangle + \triangle$$

Return the value of $\triangle$ to Ba-Gua.

---

Student's name: _____  Assignment date: _____

---

## ☷000

Chinese Magic Square (幻方) is called 纵横图，九宫数 in China.

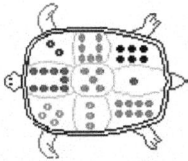

According to ancient Chinese legends, a giant tortoise surfaced from the River Lo in central China around 4,000 years ago. The ancient Chinese found a pattern on a tortoiseshell (refer to the picture). There were circular dots of numbers on the giant tortoise arranged in a three-by-three grid pattern on its shell.

The pattern of numbers on the giant tortoise in any given direction, i.e. horizontal, vertical, or diagonal, all add up to a total of 15.

Place each number from 1 to 6 in each of the following circle such that each the sum of three circles on each side of the triangle is 9.

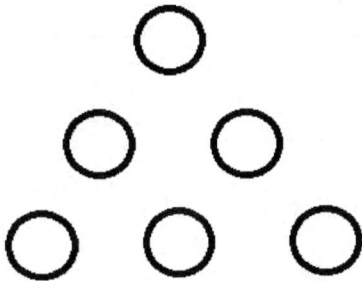

Return the value of the top circle as an answer to Ba-Gua.

## ☷001

Place each number from 1 to 6 in each of the following circle such that each the sum of three circles on each side of Triangle is 10.

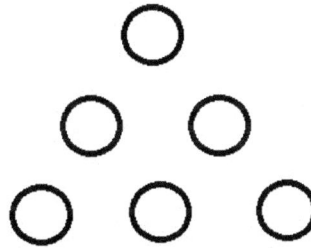

Place each number from 1 to 6 in each of the following circle such that each the sum of three circles on each side of Triangle is 11.

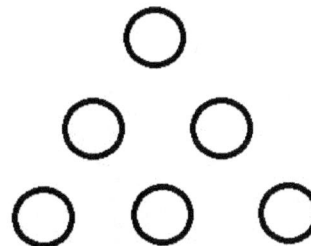

Place each number from 1 to 6 in each of the following circle such that each the sum of three circles on each side of \triangle is 12.

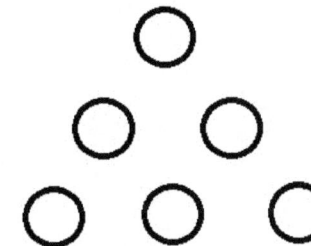

.

Return the total of the top three circle values to Ba-Gua.

---

Student's name: _____  Assignment date: _____

| ☷101 | ☳011 |
|---|---|
| Annabelle has three times as many apples as Sunny. Together they have 64 apples. How many apples does each one of them have?<br><br>Return Annabelle's answer to Ba-Gua. | Place 3, 4, 5, 6, 7, 8 in each of the following square box such that the sum of 4 numbers in each circle is 17.<br><br>Return the answer to the rightmost square box to Ba-Gua.<br><br>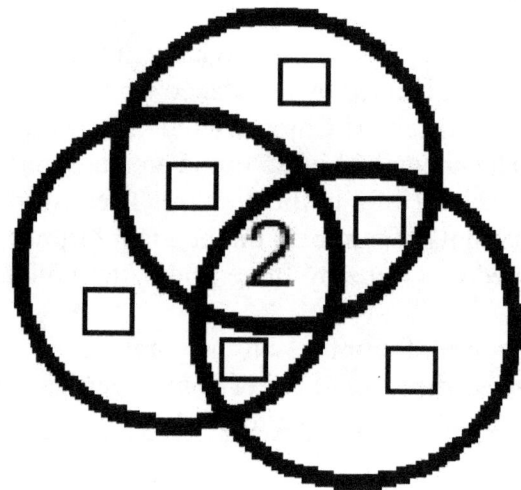 |

Student's name: _____ Assignment date: _____

**Category 11, Intelligent arithmetic worksheets by selecting calculation operator**

How can one create a calculation worksheet which allows students to decide its calculation operator? The following is the one we created to answer the above question.

Multiplication, addition, and subtraction

| 3 | 1 | 2 | 3 |
|---|---|---|---|
| 2 | 4 | 9 | 6 |
| 1 | 7 | 8 | 9 |
|   | a | b | c |

You are at b2 = ☐.

| 3 | 36 | 81 | 27 |
|---|----|----|----|
| 2 | 63 | 40 | 54 |
| 1 | 18 | 45 | 36 |
|   | d  | e  | f  |

☐ × ✥ ＋ or ━ (circle one) _____ = (shaded square on the right)

| 3 |   |   |   |
|---|---|---|---|
| 2 | x |   |   |
| 1 |   |   |   |
|   | d | e | f |

_____ × _____ ＋ or ━ (circle one) _____ = _____

9 × 4 + 27 = 63

☐ × ✥ ＋ or ━ (circle one) _____ = (shaded square on the right)

| 3 | x |   |   |
|---|---|---|---|
| 2 |   |   |   |
| 1 |   |   |   |
|   | d | e | f |

_____ × _____ ＋ or ━ (circle one) _____ = _____

Student's name: _____ Assignment date: _____

## References

(1) Ho, Frank, Vector 2006, Volume 47, Issue 2, Enriching Math Using Chess

(2) Ho, Frank, Vector 2007, Volume 48, Issue 3, A New Chess Set for Teaching Mathematical Chess

Student's name: _____ Assignment date: _____

## Category 12, Mixed 4 operations with 32 problems on one page

These are some sample four questions that use Ho Math Chess' invention of SCL (Symbolic Chess Language). These puzzle-like questions allow children to create the specific questions themselves by following SCL commands using image processing with the comparison, spatial relation, logic, and interactions etc.

| 27 | 39 | 48 |
|----|----|----|
| 29 | 3 | 58 |
| 78 | 18 | 17 |

| 12 | 21 | 31 |
|----|----|----|
| 41 | 3 | 52 |
| 62 | 71 | 82 |

| 13 | 19 | 15 |
|----|----|----|
| 18 | 3 | 14 |
| 17 | 12 | 16 |

| 12 | 21 | 18 |
|----|----|----|
| 15 | 3 | 24 |
| 6 | 9 | 27 |

$\underline{39} + \underline{3} = 42$   $\underline{21} - \underline{3} = 18$   $\underline{19} \times \underline{3} = 57$   $\underline{21} \div \underline{3} = \underline{7}$

___ + ___ = ___ 51   ___ − ___ = 9   ___ × ___ = ___ 45   ___ ÷ ___ = 4

___ + ___ = ___   ___ − ___ = ___   ___ × ___ = ___   ___ ÷ ___ = ___

___ + ___ = ___   ___ − ___ = ___   ___ × ___ = ___   ___ ÷ ___ = ___

___ + ___ = ___   ___ − ___ = ___   ___ × ___ = ___   ___ ÷ ___ = ___

___ + ___ = ___   ___ − ___ = ___   ___ × ___ = ___   ___ ÷ ___ = ___

___ + ___ = ___   ___ − ___ = ___   ___ × ___ = ___   ___ ÷ ___ = ___

___ + ___ = ___   ___ − ___ = ___   ___ × ___ = ___   ___ ÷ ___ = ___

Student's name: _____ Assignment date: _____

### Category 13, Mathematical chess puzzles

Use chess moves to solve the following puzzle.

On the first look, lots of students are not able to solve it. Why? Students are so used to do computation from left to right, and this question has to be solved in an unconventional direction.

Use chess symbol moves to solve the following puzzle.

$$\text{If } 2 \; ♖ \; 3 = 5 \text{ then } 2 \; ♗ \; 3 \text{ is } = \underline{\qquad}$$

Surprisingly, some of my students have no trouble to solve the above puzzle.

Student's name: _____ Assignment date: _____

Use chess symbol moves to solve the following puzzle.

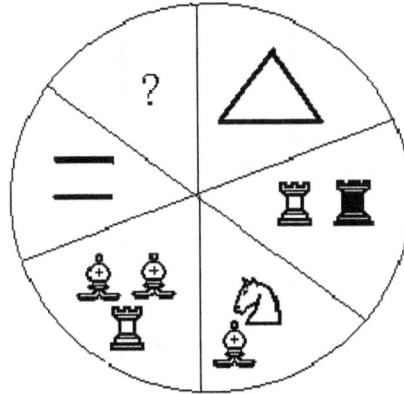

The following problem cannot be solved using chess values; students tried it and knew it. So, what is the trick behind the idea of this puzzle?

| If ♛ ÷ ♝ = ♜ |
| Then what is ♚ ÷ ♜ = ? |

Student's name: _____ Assignment date: _____

Memory and computation training

Student's name: _____ Assignment date: _____

Number relationships

The following problem shows the convergence much like to take a look at how an opponent might attack you if you make a certain move. The answer cannot be obtained if the student does not follow a sequence of events correctly and conduct multiple tasks. This training of conducting a sequence of events benefits students in real life.

| 3 | g1 | g2 | g3 |
|---|----|----|----|
| 2 | h1 | | h2 |
| 1 | i1 | i2 | i3 |
| | a | b | c |

| 3 | | | |
|---|---|---|---|
| 2 | | | |
| 1 | | | |
| | d | e | f |

| 3 | | | |
|---|---|---|---|
| 2 | | | |
| 1 | | | |
| | g | h | i |

You are at b2.

+ d1= _____ + _____ = _____

+ d2= _____ + _____ = _____

+ d3= _____ + _____ = _____

+ e1= _____ + _____ = _____

+ e2= _____ + _____ = _____

+ e3= _____ + _____ = _____

+ f1= _____ + _____ = _____

+ f2= _____ + _____ = _____

+ f3= _____ + _____ = _____

+e3+f2= ___ + ___ + ___ = _____

+d3+g2= _____ + _____ + _____ = _____

Student's name: _____ Assignment date: _____

## Category 14, Math, chess, word problems, puzzles, and Ho Math Chess specialties

I was trying to create a type of worksheets which is not just computation, not just pure word problem, or just number puzzles but a combination of all of them. Below is one of the sample worksheets, which I was trying to achieve my intention.

Question 1 is a number computation
Question 2 is a word problem.
Question 3 is a number puzzle.
The question is a number pattern.
Question 5 and 6 are famous Frankho ChessDoku and Frankho Maze.

The following is a sample.

何数棋谜连锁培训及其创始人 www.homathchess.com

Student's name: _____ Assignment date: _____

---

1. Fill in each box with a number. The same shape box has the same number.

$$\square + \square + \bigcirc + \bigcirc = 6$$
$$\square + \square + \bigcirc + \bigcirc = 8$$
$$\square + \square + \bigcirc + \bigcirc = 4$$

---

2. Each chick has 2 legs. Each piglet has 4 legs. There are 2 chicks and 20 legs altogether for all chicks and piglets. How many piglets are there?

---

3. Replace each letter with a number. The same letter means the same number.

$$\begin{array}{r} A\ B \\ +\quad B \\ \hline 8\ A \end{array}$$

---

4. Find the next number in the pattern.

8, 4, 2, _____, _____

---

5. Frankho ChessDoku™

Rule: All the digits 1 to 3 must appear exactly once in every row and column. The number that appears in the bottom right-hand corner is the result calculated according to the arithmetic operator(s) and chess move(s) as indicated by the darker arrow(s).

© 2008 Frank Ho, Amanda Ho

Ho Math and Chess

---

6. Frankho ChessMaze™

Trace the path from/to ✳□.
A darker line segment shows movement direction.

© 2008 Frank Ho, All rights reserved.

North (N)

⌄ = 1 = ♟    ✳ = 0 = ♚

---

Student's name: _____   Assignment date: _____

## Category 15, Advanced math, chess, and puzzles problems

After I accumulated more experience in creating combined math, chess, and puzzle worksheets, my difficulty level of worksheets also increased. Below is a typical worksheet that requires computation, table look-up, solving a magic square, and chess knowledge, and finally, the ability to do multi-step and multi-task. Students also need visualization ability to identify the numbers in colours.

| 5 | | 5 | | 4 | |
| 4 | 6 | | | | 3 |
| 3 | | | | | |
| | | | | | |
| 2 | 9 | | | 8 | |
| 1 | | 7 | | 6 | |
| a | b | c | d | e | |

Place each number from 2 to 10 in each of the following squares such that the sum of numbers of each row, column, or diagonal must be all equal to 18.

| red | blue | yellow |
|---|---|---|
| 11-2 | 16-8 | 12-6 |
| 14-9 | 12-7 | 17-9 |
| 13-4 | 15-6 | 15-8 |

| yellow | red | blue |
|---|---|---|
| 12-7 | 17-9 | 14-8 |
| 15-6 | 11-2 | 16-8 |
| 14-9 | 13-4 | 12-6 |

| blue | yellow | red |
|---|---|---|
| 11-2 | 17-9 | 12-7 |
| 16-8 | 15-6 | 14-5 |
| 12-6 | 13-4 | 18-9 |

(red) + = __ + __ = __

(blue) + = ___ + ___ = ___

(yellow) + = __ + __ = ___

(red) + = ___ + __ = __

(blue) + = __ + __ = ___

(yellow) + = ___ + __ = ___

(red) + = ___ + __ = ___

(blue) + = ___ + __ = ___

72

何数棋谜连锁培训及其创始人 www.homathchess.com
Frank Ho, Amanda Ho © 2015 – 2018  All rights reserved.

Student's name: _____  Assignment date: _____

**Category 16, Word problem can also be mixed with math, chess, and puzzles**

After 20 years of teaching elementary students math, the math worksheets still bug me a lot. The pure competition problems intermixed with some word problems just do not interest most students. Many high-level ability students are not challenged by doing simple "easy" word problems. For example, the following word problem will not attract highly able students at the learning centre if they have done some in their day schools.

*John has 23 pencils, and Mary has 18 pencils. How many pencils do they have altogether?*

Those word problems, which can only be solved by the top 1 to 3 students, also turn the majority of students off since they could not do them. These types of problems are better taught in the math contest class. An example is as follows:

*Melissa sold $\frac{1}{3}$ of her apples on day 1; on day 2, she sold 2 apples fewer than $\frac{2}{3}$ of the remaining apples; on day 3, she sold 3 more than $\frac{1}{4}$ of the remaining apples, and finally, she had 30 apples left. How many apples did she have originally?*

This example involves the concept of working backwards by using the quantity divided by its corresponding value to get back the unit 1 value on different days. It also involves adjusting the amount of corresponding value. If the students feel it is too abstract to solve this problem, then a drawing of a line segment diagram may help.

Student's name: _____ Assignment date: _____

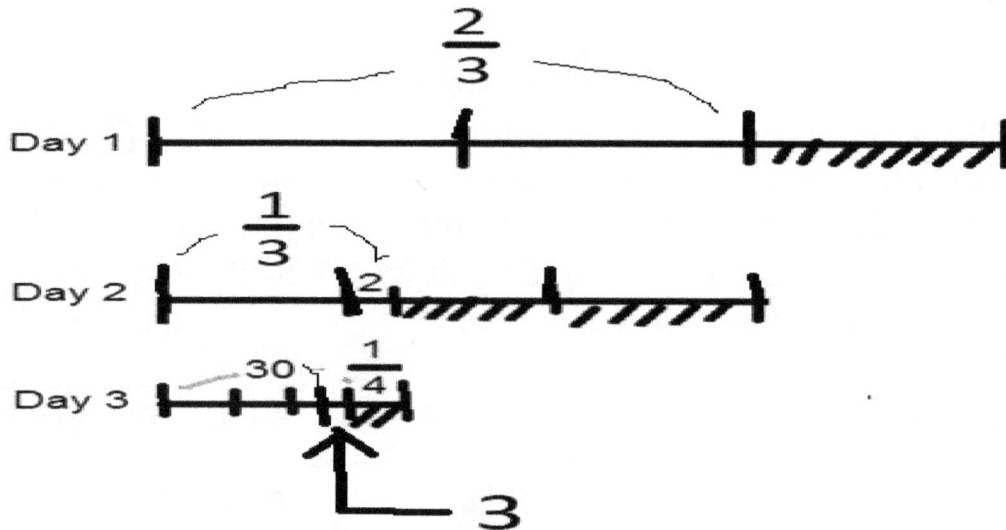

$$[(30+3) \div \frac{3}{4} - 2] \div \frac{1}{3} \div \frac{2}{3} = 189$$

When I started to give them some puzzles, especially those I created with math context in them, they showed increasing interest in working on them and even asked me for more by saying, "Can I have one of those sheets?" know I have found something.

Students like something mixed with learning and fun all in one worksheet, but this is normally done with computer software using a game approach, so how am I going to invent printable game-like or puzzle-like math worksheets? These worksheets will make the learning of mathematics more enjoyable, but this shall not be limited to just fun; otherwise, it will be just another math puzzle worksheet and will not be very productive in increasing student's math ability.

Student's name: _____   Assignment date: _____

With lots of research and observations, I have found what students like is working on some puzzles or playing chess, but I cannot just give them puzzles or let them play chess all the time in my math classes. So, what can I do? I want to invent some worksheets which use these puzzles to increase their math ability at the same time while working on a puzzle-like math worksheet that is a type of worksheet which can successfully combine math puzzles. Also, chess is what I am interested in producing.

Using the idea of integrated math, puzzles, and chess and then combined with word problems will ultimately make these types of worksheets reach the level where students can participate in the math contests if they choose to.

Ho Math Chess has been testing this type of worksheets for a few years now. The final product will be the one we discovered that seems to be more productive and just "right" for students' interests.

The following example shows how a word problem can be mixed with chess and puzzles as one subset problem in a big picture.

Student's name: _____ Assignment date: _____

---

You are a chess piece located at (e, 5).

| 6 | | | |
|---|---|---|---|
| | (a, 3) | (b, 3) | (c, 3) |
| 5 | (a, 2) | **4  1**<br>**2  3** | (c, 2) |
| 4 | | | |
| | (a, 1) | (b, 1) | (c, 1) |
| | d | E | f |

Rule: All the digits 1 to 3 must appear exactly once in every row and column.

© 2008 Frank Ho, Amanda Ho

| | a | b | c |
|---|---|---|---|
| 3 | | | |
| 2 | | | **3** |
| 1 | **1** | | |

**Ho Math Chess**

---

⊟ + The least of ⟡ = _____

⊟ + The least of ⟡ = _____

⊟ + The largest of ⟡ = _____

⊟ + The largest of ⟡ = _____

⊟ + The sum of even numbers ⟡ = __

⊟ + The sum of odd numbers ⟡ = ____

⊟ + The range of odd numbers of ⟡ = _____

⊟ + The average of the numbers of ⟡ = _____

4+2=6, 1₂₌₃
3+3=6, 2+3=5
2+2=4, 4+5=9
1+2=3, 3+7/4 = 3 3/4

Student's name: ________________ Assignment date: ___________________

Counting dots (not fingers) one-digit addition and subtraction to 10

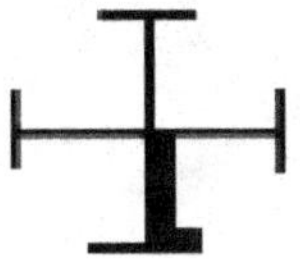

| | | | | | |
|---|---|---|---|---|---|
| 5 | 5 | 1 | 2 | 4 | 3 |
| 4 | 2 | 3 | 4 | 2 | 3 |
| 3 | 1 | 3 | 5 | 3 | 2 |
| 2 | 4 | 2 | 4 | 4 | 1 |
| 1 | 2 | 3 | 2 | 3 | 0 |
| | a | b | c | d | e |

You are at C3 = □

□ + △ = __

+△ + □ = __

○

− __

□

Check Back

△ + □ = __

+□ + △ = __

○

− __

△

Check Back

□ = ○ − △, △ = ○ − □

□ + ? < ○, ? = ____________________

○ − ? < △, ? = ____________________

○ − ? < □, ? = ____________________

△ + ? < ○, ? = ____________________

Student's name: _____ Assignment date: _____

### Category 17, Chinese and math 何中文數學

I also created a Chinese and math combined arithmetic to figure out the mathematical answer after reading a Chinese poem. All Chinese characters are square-shaped, and many of them are in symmetry.

Example 1

寫出下一個連接辭句。

上面下面
左邊右邊
左上右下

_____

寫完上面的辭句後，猜一個字 _____。

Student's name: _____ Assignment date: _____

Example 2

左一撇,右一撇,下一列
左一行,右一行,上一列,下一列

畫出下一個圖形。

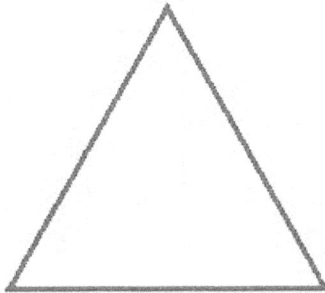

I have created about 40 problems so far, and they could be found at the following website:

https://www.scribd.com/doc/251151886/Ho-Chinese-and-Math

Student's name: _____  Assignment date: _____

**Part 3 Classification by math topics**

Example 1 - Number Concepts and Operations: Addition and Subtraction

The following problem is designed to be different from traditional worksheets, which are always from left to right or top to down in a linear fashion. One could work out the problem below from the bottom to top and then from top to down in multi-direction. It uses a chess symbol as part of the computation; thus, multi-step problems are created.

| Bottom/up and top/down operation | Bottom/up and top/down operation |
|---|---|
| | |

Example 2 - Number Concepts and Operations: Multiplication of doubling

The following problem is created with the mind that children do not learn math in a sequential way of addition, subtraction, multiplication, or division in real life. This example demonstrates a times table created using different formats. A simple multiplication problem is changed to multi-direction, multi-operation, multi-step, and multi-concept learning.

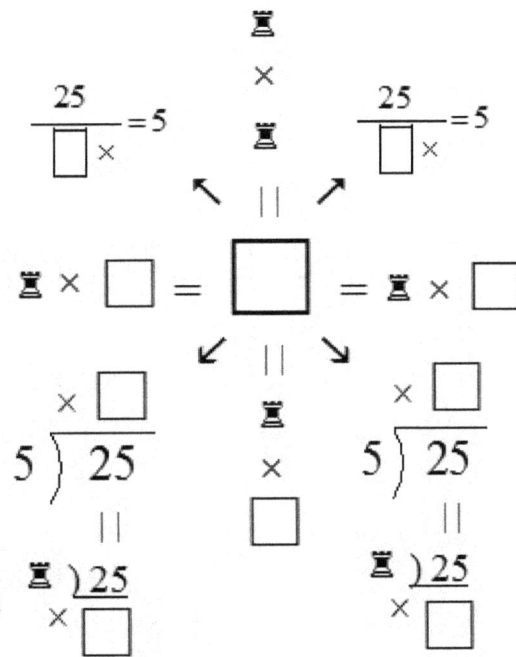

何数棋谜连锁培训及其创始人 www.homathchess.com

Frank Ho, Amanda Ho © 2015 – 2018   All rights reserved.

Student's name: _____ Assignment date: _____

Example 3 - Number Concepts and Operations: Addition and Subtraction, If Then - Else

The following operation takes a circular motion clockwise.

$$19 \quad - \quad ♚ \quad = \quad \square$$

$$
\begin{array}{r}
+ \quad\quad 9 \quad + \quad ♚ \quad = \quad \square \quad + \\
\hline
\quad \square \quad\quad\quad\quad \square
\end{array}
$$

If $10 + ♜ = \square$ , then $9 + ♜$ must be $\square$ .

If $♜ + 10 = \square$ , then $♜ + 9$ must be $\square$ .

Example 4 - Number Concepts and Operations: Cross Multiplication

♟ =          8 =
□           □

×   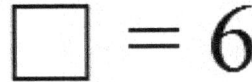   ×

□           □

□   +   □ = 6

### Example 5 - Number Concepts and Operations: Multiplication and Division

Will the following operations confuse children? The answer is no. ♛×♞ will not make sense if it is explained literally as a queen times a knight. However, if it is translated into numerals 9 times 3, then the product must be 27, which is very logical, and children understand that they are working on the product of 9 × 5, not the product of ♛×♞.

A similar type of logic question is as follows: If 2 # 3 is defined as 2 + 2 × 3, then what is 3 # 4? Normally 2 # 3 will not make any sense since it is not a valid arithmetic operator, but if we define it clearly, then it becomes workable.

♛
× 2
_____
$18 \div 2 = \square$

♛
× ♜
_____
$\square \div 9 = \square$

♛
× ♜
_____
$\square \div 5 = \square$

♛
× ♛
_____
$\square \div ♛ = \square$

♛
× ♞
_____
$\square \div ♞ = \square$

♛
× ♞
_____
$\square \div ♛ = \square$

No Part of this publication can be copied, duplicated, or reproduced.

84

Student's name: _____ Assignment date: _____

Example 6 – Pattern and Relations: Equation

The following example demonstrates how chess symbols and chess values are integrated with arithmetic operations.

$$\text{♛} + \text{♞} + x = 54$$
$$x = \underline{\hspace{2cm}}$$

Example 7 – Pattern and Relations: Pattern

Chess pieces can move horizontally, vertically, or diagonally, and the concept of symmetry is not in the learning outcome of grade K to 1, so they must thoroughly explain to students. An example of a math and chess integrated puzzle using a chess move is as follows.

Use chess moves to solve the following puzzle.

On the first look, lots of students are not able to solve it. Why? Students are so used to do computation from left to right, and this question has to be solved in an unconventional direction. Chess is a $360^0$ Visualization game, and this example demonstrates how a knight move would help solve this puzzle.

Student's name: _____ Assignment date: _____

## Example 8 Set Theory shown by two-column format

| | |
|---|---|
| Cross mark (✗) the square(s) where all rooks could share the common squares. | Find the common factors of the following numbers. |

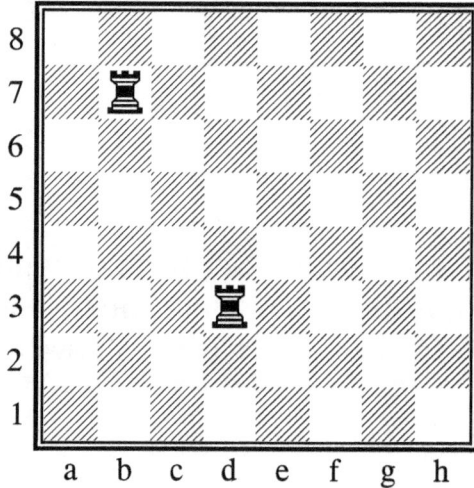

d7, b3

12, 24

13, 26

| | |
|---|---|
| Cross mark (✗) the square(s) where all rooks could share the common squares. | Find the common factors of the following numbers. |

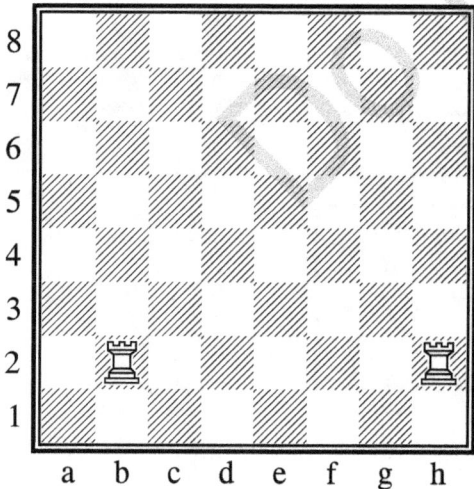

c2, d2, e2, f2, g2

11, 121

3, 26

Student's name: _____ Assignment date: _____

Example 9 An example using chess moves

| If ♛ ÷ ♝ = ♜ |
|---|
| Then what is ♚ ÷ ♜ =? |

Example 10 Shape and Space - The following is a puzzle that requires the knowledge of chess moves.

| Filling in with a chess piece | Geometric shapes |
|---|---|
| ♟ | |
| ♜ | |
| ▢ | |
| ♝ | |

Student's name: _____ Assignment date: _____

Example 11 – Pattern and Relations

Find values to replace? or fill in □. An example using chess pieces values, and logic.

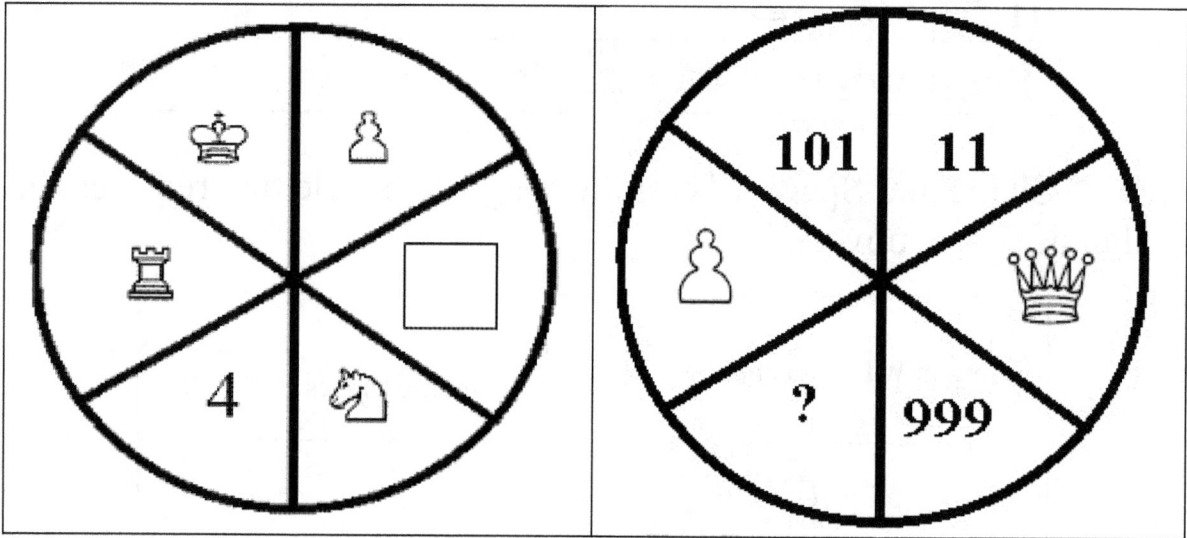

Example 12 - Use chess symbol moves to solve the following puzzle.

If 2 ♖ 3 = 5 then 2 ♗ 3 is = _____

Surprisingly, some of my students have no trouble to solve the above puzzle.

Example 13 – Shape and Space

Use chess symbol moves to solve the following puzzle.

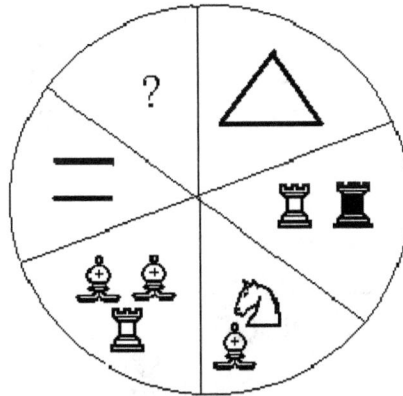

Student's name: _____ Assignment date: _____

Example 14 - Statistics and Probability

Table Values

The use of chess values is much like the use of monetary values. When children see chess or money figures, they both represent some pre-defined meaningful values. The following is an example where chess pieces' values could be monetary values, and "Total Points" could be the sum of the total money.

Fill in the different number of chess pieces to come up with each total.

| Number of ♟ | Number of ♞ | Number of ♜ | Total points |
|---|---|---|---|
| 1 | 1 | 1 | 9 |
| 3 | 2 | 0 | 9 |
| 0 | 3 | 0 | 9 |
| ☐ | ☐ | ☐ | 10 |
| ☐ | ☐ | ☐ | 10 |
| ☐ | ☐ | ☐ | 11 |
| ☐ | ☐ | ☐ | 12 |
| ☐ | ☐ | ☐ | 13 |
| ☐ | ☐ | ☐ | 14 |
| ☐ | ☐ | ☐ | 15 |

Student's name: _____   Assignment date: _____

Example 15 - Statistics and Probability

| Chess Puzzle Samples | Expected Math Learning Outcomes | Chess knowledge required |
|---|---|---|
| What is the probability of the following two pieces of meeting together? | • Probability | • Chess pieces moves |

Student's name: _____　Assignment date: _____

Example 16 - How does the knight move?

| How does a knight move? | When 2 is not = 2? |
|---|---|
| Move the knight at d5 to each square identified by ⬚. For example, it will take two moves from d5 to move to b5. Nd5 – c7 – b5.  Write the least number of moves required to reach each ⬚ from Nd5 on the squares of ⬚ | **Observe your results and see if all your answers are 2's?**<br><br>**Answer:** _____.<br><br>If each move is identified as 1 and 2 moves are 2, then you would see that in chess, the meaning of "move" takes a different meaning since not all 2-move have an equal distance from d 5. For example, it takes knight 2 equal moves from d5 to b5 and from d5 to a4, yet physically a4 is farther away from d5 than b5.<br><br>**How many moves does it take for the knight at d5 to move to the 5 White squares not identified by ⬚ on the left side diagram?**<br><br>**Answer:** _____ |

Yes, 4 moves

Student's name: _____   Assignment date: _____

Example 17 Diagonal length = Side length

Is the distance from c6 to c8 the same as distance c6 to a8?

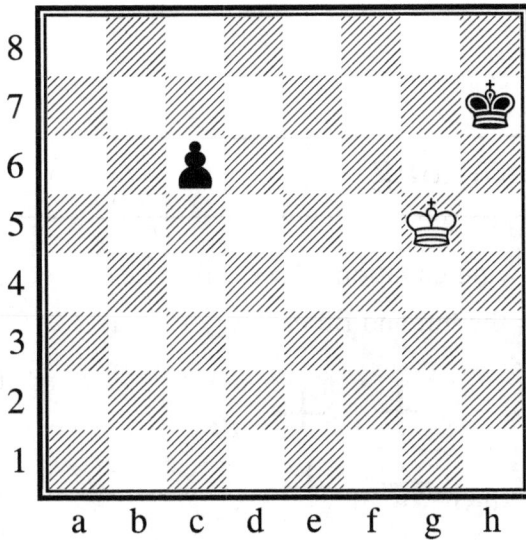

Student's name: _____   Assignment date: _____

## Example 18 - Pattern and Relations

The following problem demonstrates that children will not get confused about traditional chess symbols used in an arithmetic expression; they could use additional "creative" chess symbols and solve them correctly. This problem is suitable for grade 3 and above students.

| Chess Symbol | Logic Training |
|---|---|

**Chess Symbol**

A new Chess Symbol is defined as follows:

| Chess figurines | Chess symbols |
|---|---|
| ♔ (King) | $\div$ (Opposition) |
| ♖ (Rook) | $+$ |
| ♘ (Knight) | $\llcorner$ |
| ♗ (Bishop) | $\times$ |
| ♕ (Queen) | $*$ |
| ♙ (Pawn) | $\downarrow$ |

**Logic Training**

In the following equation, observe the chess symbols on the left and fill each $\bigcirc$ with a number.

If $+ \; + \; + \; = 10$

then $\div \; + \; + \; = \bigcirc$

$$+ \quad \begin{array}{c} + \quad * \\ + \quad \div \\ \hline \bigcirc \; \div \; * \end{array}$$

---

Use the above Chess Symbol table; find the following pattern:

Z, $\div$, O, $\downarrow$, T, $\llcorner$, T, $\times$, F, $+$, _____, $*$

Use the above Chess Symbol table; find the following pattern:

0, $\div$, 1, $\downarrow$, ___, $\llcorner$, 3, $\times$, 5, $+$, ___, $*$

**Appendix A, A List of Ho Math Chess publication**

Please contact fho1928@gmail.com for the most updated version of workbooks. Some of our workbooks are available for purchase at www.amazon.com.

- **Ultimate Math Contest Preparation, Problem Solving Strategies, and Math IQ Puzzles (3 in 1 workbook) Grade 1 and 2 (student 6717980, teacher 6720849)**
- **Ultimate Math Contest Preparation, Problem Solving Strategies, and Math IQ Puzzles (3 in 1 workbook) Grade 2 and 3 (student 6745934, teacher 6783062)**
- **Ultimate Math Contest Preparation, Problem Solving Strategies, and Math IQ Puzzles (3 in 1 workbook) Grade 3 and 4 (student 6812631, teacher 6813413)**
- **Ultimate Math Contest Preparation, Problem Solving Strategies, and Math IQ Puzzles (3 in 1 workbook) Grade 4 and 5 (student 6906686, teacher 6907410)**
- **Ultimate Math Contest Preparation, Problem Solving Strategies, and Math IQ Puzzles (3 in 1 workbook) Grade 5 and 6 (student 6979530, teacher 6985906)**
- **Ultimate Math Contest Preparation, Problem Solving Strategies, and Math IQ Puzzles (3 in 1 workbook) Grade 6 and 7 (student 7060502, teacher 7060972)**
- **Primary Grades Math (Grades 4 and under)**
- **Elementary Grades Math (Grades 5 and Up)**
- **Ho Math, Chess, and Puzzles for Grade 1 and Under**
- **Pre-K and junior kindergarten**
- **Learning Chess to Improve math**
- **Frankho ChessDoku 3 by 3**
- **Frankho ChessDoku 4 by 4**
- **Frankho ChessDoku 5 by 5**
- **Mom! I Learn Addition Using Math-Chess-Puzzles Connection**
- **Mom! I Learn Subtraction Using Math-Chess-Puzzles Connection**
- **Mom! I Learn Multiplication Using Math-Chess-Puzzles Connection**
- **Mom! I Learn Division Using Math-Chess-Puzzles Connection**

Student's name: _____ Assignment date: _____

**Appendix B Ho Math Chess Course Description**

This description is provided here for reference and may not reflect the current courses offered.

# Little Math Master for Pre-K or Kindergartener

*Pre- and Kindergarten Math*
*Problem Solving and Math IQ Puzzles for Primary Student*
*Learning chess to Improve math*
*Learning Calculation without Using Fingers*
*Frankho ChessDoku 3 by 3*

# Private School Math Entrance Test Preparation

Many parents are confused about preparing their children to prepare the math entrance test for private schools. Ho Math Chess has published math workbooks that advance children's math competition ability and increase children's math knowledge in a very systematic method. Many numbers sense strategies are taught to children to make them feel learning math computation is not challenging but fun.

The following workbooks published by Ho Math Chess could be used:

*Private School Entrance Exam for Primary Students*
*Problem Solving and Math IQ Puzzles for Primary Students*
*Frankho ChessDoku 3 by 3*
*Learning chess to Improve math*

Student's name: _____ Assignment date: _____

# Math Dyscalculia Improvement

Do you have a child who seems to have difficulties in learning very basic calculations or lacks the ability to master computation skills? This could be a sign of Dyscalculia.

Frank Ho has published a few articles on Dyscalculia, which could be found at the following website.

http://infloria.com/how-to-teach-dyscalculia-kindergarten-children/

http://www.linkroll.com/Reference-and-Education--325140-Dyscalculia-Children-Cases-Documented-Article-1.html

https://www.scribd.com/document/128605055/My-Experience-of-Teaching-Children-With-Math-Disability-or-Dyscalculia

Ho Math Chess also published a few workbooks to help students with math disabilities or dyscalculia.

*Pre-K and Kindergarten Math: Remediation for Math Dyscalculia and Math*

*Disability Learning Calculation without Using Fingers*

*Frankho ChessDoku 3 by 3*

*Learning chess to Improve math*

Student's name: _____ Assignment date: _____

# A + Math class

The fear of working on dreaded traditional computation worksheets is over. Try our world-famous fun and educational math, chess, and puzzles truly integrated and copyrighted (Canada 1069744) worksheets

Children learn math, chess, puzzles, word problems, and math contests all in one class. Ho Math Chess has been the world leader, and expert in teaching elementary math using its copyrighted and world's first math, chess, and puzzles truly integrated workbooks since 1995. By "walking through" puzzle-like worksheets, children are more focused on learning with interest through spatial relations, table look-ups, sorting, classifying, comparing, patterns, and analyzing etc. The research results of using this revolutionary teaching method have shown the statistically significant impact on raising children's math marks, improving their problem-solving skills, and improve brainpower.

A research article written by Frank Ho on Enriching Math Using Chess can be found at www.homathchess.com. The condensed version of this program has been offered at St. George's summer school for over 10 years.

Ho Math Chess has published the following workbooks to be used in the class.

- *Problem Solving and Math IQ Puzzles Volumes 1 to 5*
- *Fundamental Math*
- *Frankho 3 by 3 ChessDoku*
- *Frankho 4 by 4 ChessDoku*
- *Treasure of Math Problems*
- *Learning chess to Improve math*

何数棋谜连锁培训及其创始人 www.homathchess.com
Frank Ho, Amanda Ho © 2015 – 2018  All rights reserved.

Student's name: _____ Assignment date: _____

# Math Contest Preparation for Elementary and Middle schools

Ho Math Chess has published a few math contest preparation workbooks. Ho Math Chess is also the sponsor of the Kangaroo Math Contest.

The following Ho Math Chess workbooks could be used in the classes.

*Ultimate Math Contest Preparation - Basics*
*Ultimate Math Contest Preparation for Beginners*
*Ultimate Math Contest Preparation for Intermediate Students*
*Ultimate Math Contest Preparation for Advanced Students Volume 1*
*Ultimate Math Contest Preparation for Advanced Students Volume 2*

# SSAT Math Preparation

Ho Math Chess has published work specifically for preparing SSAT. It not only could be used to increase the SSAT percentile, but it also could increase a student's math knowledge. It is not just a test preparation workbook. It could also be used as a fast-track workbook to raise students' math level to a higher level than their peers. Ho Math Chess could offer a student a reference that could let the student stand out above all others because Ho Math Chess also teaches chess and puzzles and could train students in math contests.

The following workbooks could be used in the classes.

*SSAT Quantitative Sections Preparation for Grades 3 to 7*

## *Mini-school Math Preparation*

Ho Math Chess has published a workbook which also includes cognitive preparation. Many of our previous students have entered into Lord, Byng, Magee, Point Grey, Prince Wales and Churchill mini school or special math programs.

The following workbooks could be used in the preparation.

*Problem Solving and Math IQ Puzzles Volume 5*

*SSAT Math Quantitative Sections Preparation for Grades 3 to 7*

*Treasure of Math Problems*

## Private School Admission Application Assistance

We offer a free consultation to parents who need elementary or secondary private school admission application assistance if the student has met the following requirements:

- Must be willing to learn and to accept academic challenging.
- Must be willing to build good study habits and be willing to do homework as assigned.
- Students must be in our math program for at least 6 months in order for us to issue a recommendation letter or provide a reference so that we get to know students better. The time requirement is also to allow students to have enough time to improve their ability.

Our recommendation or reference letter can demonstrate students' extra accomplishments and math because we could offer math, chess, and puzzles integrated programs to students.

Student's name: _____  Assignment date: _____

## Chess for Juniors

### (Age 4 and up)

Chess teaching is also our specialty. Frank Ho had wanted to teach his 5-year old son how to play chess but did not know-how. So, Frank researched the game and taught chess. Frank's son then became the youngest Canadian chess master at the age of 12 and a FIDE chess master. This is how math and chess got started.

The following workbook published by Ho Math Chess will be used:

*Learning Chess to Improve Math*

www.ingramcontent.com/pod-product-compliance
Lightning Source LLC
Chambersburg PA
CBHW051416200326

41520CB00023B/7258